BIOLOGICAL AND MEDICAL PHYSICS, BIOMEDICAL ENGINEERING

For further volumes:
http://www.springer.com/series/3740

BIOLOGICAL AND MEDICAL PHYSICS, BIOMEDICAL ENGINEERING

The fields of biological and medical physics and biomedical engineering are broad, multidisciplinary and dynamic. They lie at the crossroads of frontier research in physics, biology, chemistry, and medicine. The Biological and Medical Physics, Biomedical Engineering Series is intended to be comprehensive, covering a broad range of topics important to the study of the physical, chemical and biological sciences. Its goal is to provide scientists and engineers with textbooks, monographs, and reference works to address the growing need for information.

Books in the series emphasize established and emergent areas of science including molecular, membrane, and mathematical biophysics; photosynthetic energy harvesting and conversion; information processing; physical principles of genetics; sensory communications; automata networks, neural networks, and cellular automata. Equally important will be coverage of applied aspects of biological and medical physics and biomedical engineering such as molecular electronic components and devices, biosensors, medicine, imaging, physical principles of renewable energy production, advanced prostheses, and environmental control and engineering.

Editor-in-Chief:
Elias Greenbaum, Oak Ridge National Laboratory, Oak Ridge, Tennessee, USA

Editorial Board:

Masuo Aizawa, Department of Bioengineering, Tokyo Institute of Technology, Yokohama, Japan

Olaf S. Andersen, Department of Physiology, Biophysics & Molecular Medicine, Cornell University, New York, USA

Robert H. Austin, Department of Physics, Princeton University, Princeton, New Jersey, USA

James Barber, Department of Biochemistry, Imperial College of Science, Technology and Medicine, London, England

Howard C. Berg, Department of Molecular and Cellular Biology, Harvard University, Cambridge, Massachusetts, USA

Victor Bloomfield, Department of Biochemistry, University of Minnesota, St. Paul, Minnesota, USA

Robert Callender, Department of Biochemistry, Albert Einstein College of Medicine, Bronx, New York, USA

Steven Chu, Lawrence Berkeley National Laboratory, Berkeley, California, USA

Louis J. DeFelice, Department of Pharmacology, Vanderbilt University, Nashville, Tennessee, USA

Johann Deisenhofer, Howard Hughes Medical Institute, The University of Texas, Dallas, Texas, USA

George Feher, Department of Physics, University of California, San Diego, La Jolla, California, USA

Hans Frauenfelder, Los Alamos National Laboratory, Los Alamos, New Mexico, USA

Ivar Giaever, Rensselaer Polytechnic Institute, Troy, New York, USA

Sol M. Gruner, Cornell University, Ithaca, New York, USA

Judith Herzfeld, Department of Chemistry, Brandeis University, Waltham, Massachusetts, USA

Mark S. Humayun, Doheny Eye Institute, Los Angeles, California, USA

Pierre Joliot, Institute de Biologie Physico-Chimique, Fondation Edmond de Rothschild, Paris, France

Lajos Keszthelyi, Institute of Biophysics, Hungarian Academy of Sciences, Szeged, Hungary

Robert S. Knox, Department of Physics and Astronomy, University of Rochester, Rochester, New York, USA

Aaron Lewis, Department of Applied Physics, Hebrew University, Jerusalem, Israel

Stuart M. Lindsay, Department of Physics and Astronomy, Arizona State University, Tempe, Arizona, USA

David Mauzerall, Rockefeller University, New York, New York, USA

Eugenie V. Mielczarek, Department of Physics and Astronomy, George Mason University, Fairfax, Virginia, USA

Markolf Niemz, Medical Faculty Mannheim, University of Heidelberg, Mannheim, Germany

V. Adrian Parsegian, Physical Science Laboratory, National Institutes of Health, Bethesda, Maryland, USA

Linda S. Powers, University of Arizona, Tucson, Arizona, USA

Earl W. Prohofsky, Department of Physics, Purdue University, West Lafayette, Indiana, USA

Andrew Rubin, Department of Biophysics, Moscow State University, Moscow, Russia

Michael Seibert, National Renewable Energy Laboratory, Golden, Colorado, USA

David Thomas, Department of Biochemistry, University of Minnesota Medical School, Minneapolis, Minnesota, USA

P. Gruber · D. Bruckner
C. Hellmich · H.-B. Schmiedmayer
H. Stachelberger · I.C. Gebeshuber
Editors

Biomimetics – Materials, Structures and Processes

Examples, Ideas and Case Studies

With 122 Figures

Editors

Dipl.Ing. Dr.techn. Petra Gruber
Zentagasse 38/1
1050 Wien, Austria
E-mail:
peg@transarch.org

Dr. Dietmar Bruckner
Technische Universität Wien
Institut für Computertechnik
Gusshausstr. 27–29
1040 Wien, Austria
E-mail:
bruckner@ict.tuwien.ac.at

Professor Dr. Christian Hellmich
Technische Universität Wien
Institut für Mechanik
der Werkstoffe und Strukturen
Karlsplatz 13
1040 Wien, Austria
E-mail:
christian.hellmich@tuwien.ac.at

Dr. Heinz-Bodo Schmiedmayer
Technische Universität Wien
Institut für Hochbau und Technologie
Wiedner Hauptstr. 8–10
1040 Wien, Austria
E-mail:
Heinz-Bodo.Schmiedmayer@tuwien.ac.at

Professor Dr. Herbert Stachelberger
Technische Universität Wien, USTEM
Getreidemarkt 9
1060 Wien, Austria
E-mail:
stachelberger@ustem.tuwien.ac.at

Professor Dr. Ille C. Gebeshuber
Technische Universität Wien
Institut für Angewandte Physik
Wiedner Hauptstr. 8–10/134
1040 Wien, Austria
E-mail:
gebeshuber@iap.tuwien.ac.at

Biological and Medical Physics, Biomedical Engineering   ISSN 1618-7210
ISBN 978-3-642-11933-0   e-ISBN 978-3-642-11934-7
DOI 10.1007/978-3-642-11934-7
Springer Heidelberg Dordrecht London New York

Library of Congress Control Number: 2011929499

© Springer-Verlag Berlin Heidelberg 2011
This work is subject to copyright. All rights are reserved, whether the whole or part of the material is concerned, specifically the rights of translation, reprinting, reuse of illustrations, recitation, broadcasting, reproduction on microfilm or in any other way, and storage in data banks. Duplication of this publication or parts thereof is permitted only under the provisions of the German Copyright Law of September 9, 1965, in its current version, and permission for use must always be obtained from Springer. Violations are liable to prosecution under the German Copyright Law.
The use of general descriptive names, registered names, trademarks, etc. in this publication does not imply, even in the absence of a specific statement, that such names are exempt from the relevant protective laws and regulations and therefore free for general use.

Cover design: eStudio Calamar Steinen

Printed on acid-free paper

Springer is part of Springer Science+Business Media (www.spinger.com)

Preface

It is said very often that humankind should learn from nature. This means that some sort of technology transfer from biology to engineering has to be established. Nowadays, terms such as bionics, biomimetics, or bio-inspiration have been introduced to describe the concepts by which ideas of technology are derived from nature. One of the most important insights so far is that it is not feasible to try to copy nature. As many examples have shown, it makes sense to start with a careful analysis and abstraction of biological processes and structures. The implementation process itself requires substantial adaptation using common engineering knowledge to guarantee successful solutions. It is not surprising that in the field of biomimetics/bionics, principles of evolution or strategies of evolution have gained much attention. The primary goal often lies in surpassing "Nature" and thus achieving outstanding results. All these aspects are to a considerable extent taken into account and covered by the various topics of the contributions in this book. The main aim of this book is therefore to provide the reader with essential information on how biomimetic/bionic working principles are identified and also brought to technical implementation in various engineering disciplines.

Vienna *Petra Gruber, Dietmar Bruckner*
June 2011 *Christian Hellmich, Heinz-Bodo Schmiedmayer*
Herbert Stachelberger, Ille C. Gebeshuber

Contents

1 **Biomimetics: Its Technological and Societal Potential** 1
   Herbert Stachelberger, Petra Gruber, and Ille C. Gebeshuber

Part I Material Structure

2 **Bionic (Nano) Membranes** 9
   Jovan Matovic and Zoran Jakšić
   2.1   Artificial Nanomembranes 10
   2.2   Biological Nanomembranes 12
   2.3   Functionalization of Artificial Nanomembranes
         Toward Bionic Structures at ISAS: TU Wien 13
         2.3.1   Nanomembrane-Based Bionic Structures
                 for Energy Harvesting 13
         2.3.2   Nanomembranes as Bionic Detectors
                 of Electromagnetic Radiation 19
   2.4   Conclusion 22

3 **Biomimetics in Tribology** 25
   I.C. Gebeshuber, B.Y. Majlis, and H. Stachelberger
   3.1   Introduction: Historical Background and Current
         Developments 26
   3.2   Biology for Engineers 28
   3.3   Method: The Biomimicry Innovation Method 30
   3.4   Results: Biomimetics in Tribology – Best Practices
         and Possible Applications 32
         3.4.1   Application of the Biomimicry Innovation
                 Method Concerning Mechanical Wear 33
         3.4.2   Application of the Biomimicry Innovation
                 Method Concerning Shear 35

		3.4.3	Application of the Biomimicry Innovation Method Concerning Tension	35
		3.4.4	Application of the Biomimicry Innovation Method Concerning Buckling, Fatigue, Fracture (Rupture) and Deformation	35
		3.4.5	Application of the Biomimicry Innovation Method Concerning Attachment	37
	3.5	Summary and Outlook		40
4	**Reptilian Skin as a Biomimetic Analogue for the Design of Deterministic Tribosurfaces**			**51**
	H.A. Abdel-Aal and M. El Mansori			
	4.1	Introduction		52
	4.2	Background		56
		4.2.1	The Python Species	56
		4.2.2	Structure of Snake Skin	58
		4.2.3	Skin Shedding	59
	4.3	Observation of Shed Skin		60
		4.3.1	Initial Observations	60
		4.3.2	Optical Microscopy Observations	62
		4.3.3	Scan Electron Microscopy Observations	63
	4.4	Metrology of the Surface		69
		4.4.1	Topographical Metrology	69
		4.4.2	Bearing Curve Analysis	70
	4.5	Correlation to Honed Surfaces		73
	4.6	Conclusions and Future Outlook		77
5	**Multiscale Homogenization Theory: An Analysis Tool for Revealing Mechanical Design Principles in Bone and Bone Replacement Materials**			**81**
	Christian Hellmich, Andreas Fritsch, and Luc Dormieux			
	5.1	Introduction		84
	5.2	Fundamentals of Continuum Micromechanics		85
		5.2.1	Representative Volume Element	85
		5.2.2	Upscaling of Elasto-Brittle and Elastoplastic Material Properties	86
	5.3	Bone's Hierarchical Organization		88
	5.4	Elastic and Strength Properties of the Elementary Components of Bone: Hydroxyapatite, Collagen, Water		88
	5.5	Multiscale Micromechanical Representation of Bone		91
	5.6	Experimental Validation of Multiscale Micromechanics Theory for Bone		93
	5.7	How Bone Works: Mechanical Design Characteristics of Bone Revealed Through Multiscale Micromechanics		96
	5.8	Some Conclusions from a Biological Viewpoint		98

Contents ix

6 Bioinspired Cellular Structures: Additive Manufacturing and Mechanical Properties ... 105
J. Stampfl, H.E. Pettermann, and R. Liska
- 6.1 Introduction ... 105
- 6.2 Fabrication of Bioinspired Cellular Solids Using Lithography-Based Additive Manufacturing ... 107
  - 6.2.1 Laser-Based Stereolithography ... 108
  - 6.2.2 Dynamic Mask-Based Stereolithography ... 108
  - 6.2.3 Inkjet-Based Systems ... 110
  - 6.2.4 Two-Photon Polymerization ... 111
- 6.3 Photopolymers for Additive Manufacturing Technologies ... 112
  - 6.3.1 Principles of Photopolymerization ... 112
  - 6.3.2 Radical and Cationic Systems in Lithography-Based AMT ... 114
  - 6.3.3 Biomimetic, Biocompatible, and Biodegradable Formulations ... 115
- 6.4 Mechanical Properties: Modeling and Simulation ... 118
  - 6.4.1 Linear Elastic Behavior ... 118
  - 6.4.2 Nonlinear Response ... 119
  - 6.4.3 Sample Size and Effective Behavior ... 119
- 6.5 Conclusion ... 121

Part II Form and Construction

7 Biomimetics in Architecture [Architekturbionik] ... 127
Petra Gruber
- 7.1 Introduction ... 127
- 7.2 History: Different Approaches ... 128
  - 7.2.1 Analogy and Convergence ... 129
  - 7.2.2 Strategic Search for the Overlaps Between Architecture and Nature ... 130
- 7.3 Strategies: What is Transferred and How is it Done? ... 131
  - 7.3.1 What is Transferred? ... 131
  - 7.3.2 Methods ... 131
- 7.4 Application Fields: Successful Examples ... 134
  - 7.4.1 Emergence and Differentiation: Morphogenesis ... 134
  - 7.4.2 Interactivity ... 135
  - 7.4.3 Dynamic Shape ... 135
  - 7.4.4 Intelligence ... 136
  - 7.4.5 Energy Efficiency ... 136
  - 7.4.6 Material/Structure/Surface ... 137
  - 7.4.7 Integration ... 137
- 7.5 Case Studies ... 138
  - 7.5.1 Biomimetics Design Exercise ... 138

		7.5.2	Biomimetics Design Programmes, Workshops	
			and Studies	140
	7.6	Future Fields, Aims and Conclusion		144
		7.6.1	Aims	144
		7.6.2	Considerations About Future Developments	145

8 Biomorphism in Architecture: Speculations on Growth and Form ... 149
Dörte Kuhlmann

	8.1	Introduction	149
	8.2	The Essence of Nature	150
	8.3	Nature as a Source for Form	152
	8.4	Natural Processes	153
	8.5	Organic Versus "Mechanical" Form	156
	8.6	Bionics and Cyborgs	159
	8.7	Ecology	162
	8.8	From Fractals to Catastrophies	165
	8.9	Form Follows Function	167
	8.10	The Concept of Organic Unity	171
	8.11	Conclusion	174

9 Fractal Geometry of Architecture ... 179
Wolfgang E. Lorenz

	9.1	Fractal Concepts in Nature and Architecture	179
		9.1.1 From the Language of Fractals to Classification	179
	9.2	Fractals: A Definition from a Mathematical and an Architectural Point of View	182
		9.2.1 Roughness and Length Measurement	182
		9.2.2 Scale Range and Distance	184
		9.2.3 Self-Similarity: An Important Attribute of Fractals	184
		9.2.4 Architectural Examples	186
		9.2.5 Developed Through Iteration	187
		9.2.6 Differences Between Architectural and Mathematical Fractals	189
		9.2.7 Fractals as a Design Aid	189
		9.2.8 Fractals Are Common to Nature	190
		9.2.9 The Factor Chance	191
	9.3	From Simulation to Measurement	192
		9.3.1 Curdling	192
		9.3.2 Fractal Dimension	194
		9.3.3 Perception and Distance	196
	9.4	Fractal Dimension and Architecture	196
		9.4.1 Fractal Dimension and Approaching a Building	197
		9.4.2 Results of Measurement	198
	9.5	Conclusions and Outlook	199

Part III Information and Dynamics

10 Biomimetics in Intelligent Sensor and Actuator Automation Systems 203
Dietmar Bruckner, Dietmar Dietrich, Gerhard Zucker, and Brit Müller
- 10.1 Research Field 204
- 10.2 Automation 204
- 10.3 Intelligence and Communication 206
- 10.4 Open Problems: Challenges in Research 207
- 10.5 Intelligence of Bionic Systems 209
  - 10.5.1 Hierarchical Model Conception 209
  - 10.5.2 Statistical Methods 210
  - 10.5.3 Definition of Intelligence 211
  - 10.5.4 Choice of the Right Model 212
  - 10.5.5 Top-Down Methodology 212
  - 10.5.6 A Unitary Model 213
  - 10.5.7 Differentiation Between Function, Behavior, and Projection 213
  - 10.5.8 Indispensible Interdisciplinarity 214
- 10.6 The Psychoanalytical Model 214
- 10.7 Conclusion 217

11 Technical Rebuilding of Movement Function Using Functional Electrical Stimulation 219
Margit Gföhler
- 11.1 Introduction 219
- 11.2 Principle 220
- 11.3 Actuation 220
  - 11.3.1 Stimulation Signal 222
  - 11.3.2 Electrodes 223
- 11.4 Stimulators 224
- 11.5 Control 225
  - 11.5.1 Modeling/Simulation 225
  - 11.5.2 Control Systems 227
- 11.6 Sensors 229
  - 11.6.1 Artificial Sensors 229
  - 11.6.2 Natural Sensors in the Peripheral Nervous System 229
  - 11.6.3 Volitional Biological Signals 230
- 11.7 Applications for the Lower Limb 231
  - 11.7.1 Cycling 231
  - 11.7.2 Rowing 239
  - 11.7.3 Gait 241
- 11.8 Applications for the Upper Limb 242
- 11.9 Outlook 243
- References 244

12 Improving Hearing Performance Using Natural Auditory Coding Strategies 249
Frank Rattay
- 12.1 The Hair Cell Transforms Mechanical into Neural Signals 249
- 12.2 The Human Ear 251
- 12.3 Place Theory Versus Temporal Theory 253
- 12.4 Noise-Enhanced Auditory Information 253
- 12.5 Auditory Neural Network Sensitivity Can be Tested with Artificial Neural Networks 257
- 12.6 Cochlear Implants Versus Natural Hearing 258
- 12.7 Discussion 259
- 12.8 Conclusion 260
- References 260

Index 263

Contributors

Hisham A. Abdel-Aal, Laboratoire de Mécanique et Procédé de Fabrication (LMPF, EA4106), Arts et Métiers Paris Tech., Rue St Dominique, BP 508, 51006 Châlons-en-Champagne Cedex, France, hisham.abdel-aal@chalons.ensam.fr

Dietmar Bruckner, Institute of Computer Technology, Vienna University of Technology (TU Wien), Gußhausstraße 27-29, 1040 Vienna, Austria, bruckner@ict.tuwien.ac.at

Dietmar Dietrich, Institute of Computer Technology, Vienna University of Technology (TU Wien), Gußhausstraße 27-29, 1040 Vienna, Austria, dietrich@ict.tuwien.ac.at

Luc Dormieux, Ecole des Ponts ParisTech, 6-8 av. Blaise Pascal, 77455 Marne-la-Vallée, France, luc.dormieux@enpc.fr

Andreas Fritsch, Institute for Mechanics of Materials and Structures, Vienna University of Technology (TU Wien), Karlsplatz 13/202, 1040 Vienna, Austria, andreas.fritsch@tuwien.ac.at

Ille C. Gebeshuber, Institute of Microengineering and Nanoelectronics (IMEN), Universiti Kebangsaan Malaysia, 43600 UKM, Bangi, Selangor, Malaysia
and
Institute of Applied Physics, Vienna University of Technology (TU Wien), Wiedner Hauptstrasse 8-10/134, 1040 Vienna, Austria
and
AC^{2}T Austrian Center of Competence for Tribology, Viktor Kaplan-Straße 2, 2700 Wiener Neustadt, Austria, gebeshuber@iap.tuwien.ac.at

Margit Gföhler, Research Group for Machine Elements and Rehabilitation Engineering, Institute of Engineering Design and Logistics Engineering, Vienna University of Technology (TU Wien), Getreidemarkt 9/307-3, 1060 Vienna, Austria, margit.gfoehler@tuwien.ac.at

Petra Gruber, transarch, office for biomimetics and transdisciplinary architecture, Zentagasse 38/1, 1050 Vienna, Austria, peg@transarch.org
and
Institute for History of Art and Architecture, Building Archaeology and Preservation, Vienna University of Technology (TU Wien), Karlsplatz 13 / 251-1, 1040 Vienna, Austria, petra.gruber@tuwien.ac.at

Christian Hellmich, Institute for Mechanics of Materials and Structures, Vienna University of Technology (TU Wien), Karlsplatz 13/202, 1040 Vienna, Austria, christian.hellmich@tuwien.ac.at

Zoran Jakšić, Department of Microelectronic Technologies and Single Crystals, Institute of Chemistry, Technology and Metallurgy, University of Belgrade, Belgrade, Serbia, jaksa@nanosys.ihtm.bg.ac.rs

Dörte Kuhlmann, Department of Architecture Theory E259.4, Institute of Architectural Sciences, Vienna University of Technology (TU Wien), Wiedner Hauptstr. 7, 1040 Vienna, Austria, kuhlmann@a-theory.tuwien.ac.at

Robert Liska, Institute of Applied Synthetic Chemistry, Vienna University of Technology (TU Wien), Getreidemarkt 9, 1060 Vienna, Austria, robert.liska@tuwien.ac.at

Wolfgang E. Lorenz, Digital Architecture and Planning (IEMAR) E259.1, Institute of Architectural Sciences, Vienna University of Technology (TU Wien), Treitlstraße 3/1st floor, 1040 Vienna, Austria, lorenz@iemar.tuwien.ac.at

B.Y. Majlis, Institute of Microengineering and Nanoelectronics (IMEN), Universiti Kebangsaan Malaysia, 43600 UKM, Bangi, Selangor, Malaysia, burhan@vlsi.eng.ukm.my

M. El Mansori, Arts et Métier ParisTech, Rue Saint Dominique BP 508, 51006 Chalons-en-Champagne, France, Mohamed.ELMANSORI@ensam.eu

Jovan Matovic, Institute for Sensor and Actuator Systems, Vienna University of Technology (TU Wien), Vienna, Austria, Jovan.Matovic@tuwien.ac.at

Brit Müller, Institute of Computer Technology, Vienna University of Technology (TU Wien), Gußhausstraße 27–29, 1040 Vienna, Austria, mueller@ict.tuwien.ac.at

Heinz E. Pettermann, Institute of Lightweight Design and Structural Biomechanics, Vienna University of Technology (TU Wien), Gußhausstraße 25–29, 1040 Vienna, Austria, heinz.pettermann@tuwien.ac.at

Frank Rattay, Institute for Analysis and Scientific Computing, Vienna University of Technology (TU Wien), Wiedner Hauptstrasse 8–10, Vienna, Austria, frank.rattay@tuwien.ac.at

Heinz-Bodo Schmiedmayer, Institute of Mechanics and Mechatronics, Vienna University of Technology (TU Wien), Wiedner Hauptstr. 8, 1040 Vienna, Austria, heinz-bodo.schmiedmayer@tuwien.ac.at

Herbert Stachelberger, Institute of Chemical Engineering and University Service-Center for Transmission Electron Microscopy, Vienna University of Technology (TU Wien), Getreidemarkt 9/166, 1060 Vienna, Austria
and
TU BIONIK Center of Excellence for Biomimetics, Vienna University of Technology (TU Wien), Getreidemarkt 9/134, 1060 Vienna, Austria,
hstachel@mail.zserv.tuwien.ac.at

Jürgen Stampfl, Institute of Materials Science and Technology, Vienna University of Technology (TU Wien), Favoritenstraße 9, 1040 Vienna, Austria,
jstampfl@pop.tuwien.ac.at

Gerhard Zucker, Energy Department, Austrian Institute of Technology, Giefinggasse 2, 1210 Vienna, Austria, gerhard.zucker@ait.ac.at

TU BIONIK Center of Excellence for Biomimetics, Vienna University of Technology (TU Wien), Getreidemarkt 9/134, 1060 Vienna, Austria,
bionik.tuwien.ac.at, biomimetics.tuwien.ac.at

Chapter 1
Biomimetics: Its Technological and Societal Potential

Herbert Stachelberger, Petra Gruber, and Ille C. Gebeshuber

Abstract This introductory chapter contains a short discussion of the topic of biomimetics with special emphasis on background and goals together with an overview of the book. Biomimetics is described as information transfer from biology to the engineering sciences. Methods and preconditions for this interdisciplinary scientific subject are mentioned briefly focusing on the educational issues and the pathway to product development. To provide the reader with a preliminary information, an overview of the book is given devoted to a brief description of the remaining chapters which are allocated to three main sections "Material & Structure", "Form & Construction", and "Information & Dynamics".

The process of evolution on earth during the last approximately 3.4 billion years resulted in a vast variety of living structures. Most recent findings suggest that multicellular organisms could have been around for 2.1 billion years [1, 2]. At any time, organisms were able to adapt dynamically to various environmental conditions. It is therefore the principal goal of biomimetics to provide an in-depth understanding of the solutions and strategies having evolved over time and their possible implementation into technological practice. Very often biomimetics must reach down to the microscopic and ultimately to the molecular scale. Some of nature's best tricks are conceptually simple and easy to rationalize in physical or engineering terms, but realizing them requires machinery of exquisite delicacy [3].

The routes of technology transfer from biology to the engineering sciences are normally not too clearly drawn. There is no doubt about the outstanding innovative

---

H. Stachelberger (✉)
Institute of Chemical Engineering and University Service-Center for Transmission Electron Microscopy, Vienna University of Technology, Getreidemarkt 9/166, 1060 Vienna, Austria
and
TU BIONIK Center of Excellence for Biomimetics, Vienna University of Technology, Getreidemarkt 9/134, 1060 Vienna, Austria
e-mail: hstachel@mail.zserv.tuwien.ac.at

potential of the biomimetic approach. Yet there is no guarantee that a technical solution based on biomimetics will be ecofriendly. This has to be proven separately in any case.

There is a tremendous scope of research topics in the field of biomimetics (bionics) that – according to Rick [4] – could be roughly assigned to either construction bionics (e.g. materials, prosthetics, and robotics), procedural bionics (e.g. climate/energy, building, sensors, and kinematics/dynamics), or information bionics (e.g. neurobionics, evolution bionics, process bionics, and organization bionics).

Biomimetics is therefore an innovation method being applied in a multitude of technological fields. The realization of biomimetic innovation can be done either from the technological point of view (problem-oriented) or from the biology point of view (solution-oriented). These basically different approaches are therefore called top-down (biomimetics by analogy) or bottom-up (biomimetics by induction) approaches. They can also be distinguished by their differing time of development and other requirements.

To some extent biomimetics as an interdisciplinary scientific subject is thought to contribute to sustainable innovation [5]. Complex systems and patterns arise out of a multiplicity of relatively simple interactions in a hierarchically structured world. This phenomenon is called *emergence*. According to systems theory, it is necessary to go well beyond the frontiers of classical disciplines, thought patterns, and organizational structures in order to accomplish sustainable innovation.

Comprehensive knowledge as an asset is one of the most important preconditions for innovation within knowledgeable societies (Lane 1966 cited by Jursic [6]). Applied curiosity about everything's working principles requires a profound preoccupation with the technical foundations. Biomimetics is thought to facilitate the approach to technological developments and to foster scientific basics.

Advanced training in the fields of interdisciplinary research and development is a challenge that has to be met using novel concepts of teaching and training. Training is therefore a key to the expansion of biomimetics. It should be included in the training syllabus of engineers and designers to make them aware of the potential of the approach. The biological sciences should be made aware of the commercial applications of their knowledge. In order to introduce innovation principles into societal practice, there is need for ingenious and well-educated people and a proactive environment. For the education of highly qualified scientists and engineers, open access to scientific fields and domains is indispensable. Interdisciplinary activities in research units such as universities need to be initiated and supported internally (executive level) and externally (research grants). Close cooperation is necessary between R&D units and industry and economy in order to promote inventions. The formation of biomimetics networks currently taking place in Europe can be seen as the core event in the formation of a dynamic developmental area [7, 8].

The main aim of this book is to provide the reader with a collection of chapters that review the actual R&D activities at Vienna University of Technology with respect to topics in biomimetics. The three main sections "Material & Structure",

"Form & Construction", and "Information & Dynamics" cover a wide range of topics.

"Material & Structure" contains five chapters. Matovic and Jakšić start this section with a chapter on "Bionic (Nano)Membranes" [9]. The authors offer a concise and clear picture of the most important artificial nanomembrane-related procedures and technologies, including those for fabrication and functionalization, and present the main properties and potential applications, emphasizing recent results in the field contributed by the authors. Bionic nanomembranes have a potential to improve environmental protection, to bring breakthroughs in life science, to enable the production of clean energy, and to contribute in numerous other ways to an enrichment of the overall quality of life.

Tribology, the science of friction, adhesion, lubrication, and wear, is the focus of the next two chapters. Tribology is omnipresent in biology, and various biological systems have impressive tribological properties. In "Biomimetics in Tribology" [10], Gebeshuber, Majlis, and Stachelberger investigate a large hitherto unexplored body of knowledge in biology publications that deals with lubrication and wear, but that has not before been linked extensively to technology. Best practices presented comprise materials and structures in organisms as diverse as kelp, banana leafs, rattan, diatoms, and giraffes.

In "Reptilian Skin as a Biomimetic Analogue for Design of Deterministic Tribo Surfaces" [11], Abdel-Aal and El Mansori investigate the multiscale structural features of reptilian skin. Shed skin of a Ball Python is chosen as the bioanalogue. Snakes have surface features that contribute to excellent wear resistance and tunable frictional response in demanding environments. The results are translated to enhance the textural design of cylinder liners in internal combustion engines.

Hellmich, Fritsch, and Dormieux subsequently investigate multiscale homogenization theory, an analysis tool for revealing mechanical design principles in bone and bone replacement materials [12]. Multiscale poromechanics recently became a key tool to understand "building plans" inherent to entire material classes. In bone materials, the elementary component "collagen" induces, right at the nanolevel, the elastic anisotropy, while water layers between stiff and strong hydroxyapatite crystals govern the inelastic behavior of the nanocomposite, unless the "collagen reinforcement" breaks. Mimicking this design principle may hold great potential for novel biomedical materials and for other engineering problems requiring strong and light materials.

In the final chapter in the section on materials and structure, Stampfl, Pettermann, and Liska report on "Bio-inspired cellular structures: Additive manufacturing and mechanical properties" [13]. Many biological materials (wood, bone, etc.) are based on cellular architecture. This design approach allows nature to fabricate materials that are light, but still stiff and strong. Using finite element modeling in combination with additive manufacturing, it is now possible to study the mechanical properties of such cellular structures from a theoretical and experimental point of view. Stampfl, Pettermann, and Liska give an overview of currently available additive manufacturing technologies, with a focus on lithography-based systems.

Additionally, numerical methods for the prediction of mechanical properties of cellular solids with defined architecture are presented.

The section on "Form & Construction" contains three chapters. Gruber introduces the emerging field "Biomimetics in Architecture" [14] and presents various case studies that exemplify the innovational potential of structures, materials, and processes in biology for architecture and emphasize the importance of the creation of visions with the strength to establish innovation for the improvement of the quality of our built environment.

Kuhlmann deals in her chapter "Biomorphism in Architecture – Speculations on Growth and Form" [15] with the essence of nature, nature as source for form and ecology, touches upon cyborgs and the concept of organic unity, and reaches the conclusion that despite many authors' claims of producing something radically new, many of the design strategies applied by the current architectural avant-garde can be traced back to one of the oldest and most influential ideas in architectural history: the concept of organicism in its various guises.

In the final chapter of this section, "Fractal Geometry of Architecture – Fractal Dimension as a Connection Between Fractal Geometry and Architecture" [16], Lorenz introduces fractal concepts in nature and architecture, and defines them from mathematical and architectural points of view. The fractal concept of architecture means that details of different sizes are kept together by a central rule or idea: avoiding monotony by using variation. The author concludes that this concept is the reason why Gothic cathedrals and examples of the so-called organic architecture are so interesting and diversified.

The three chapters in the section "Information & Dynamics" deal with sensors and actuators, electrostimulation of muscles, and improved strategies for auditory coding in cochlear implants. Automation, dealing with the utilization of control systems and information technology to reduce the required human intervention, mostly in industrial processing systems, but more in all kinds of daily human activities for instance driving a car in the near future, faces the problem of increasing complexity through the incorporation of dramatically increasing amounts of details in sensory systems.

In "Biomimetics in Intelligent Sensor and Actuator Automation Systems" [17], Bruckner, Dietrich, Zucker, and Müller present an approach to use biomimetics as the promising method for overcoming this problem. They argue for careful application of the biomimetics approach in various respects in order to develop a technically feasible model of the human psyche and hence redeeming one of the big promises of artificial intelligence from the early days on.

The next chapter is "Technical Rebuilding of Movement Function Using Functional Electrostimulation" by Gföhler [18]. To rebuild lost movement functions, neuroprostheses based on functional electrical stimulation (FES) artificially activate skeletal muscles in corresponding sequences, using both residual body functions and artificial signals for control. Besides the functional gain, FES training also brings physiological and psychological benefits for spinal-cord-injured subjects. Current stimulation technology and the main components of FES-based neuroprostheses including enhanced control systems are presented. Technology and application of

1 Biomimetics: Its Technological and Societal Potential

FES cycling and rowing, both approaches that enable spinal-cord-injured subjects to participate in mainstream activities and improve their health and fitness by exercising like able-bodied subjects, are discussed in detail and an overview of neuroprostheses that aim at restoring movement functions for daily life as walking or grasping is given.

In the final chapter of this book, "Improving Hearing Performance Utilizing Natural Auditory Coding Strategies", Rattay deals with cochlear implants, the most successfully applied neural prostheses [19]. Cochlear implants have still deficits in comparison with normal hearing. The author argues that in contrast to nature, spiking patterns generated artificially in the auditory nerve via actual implants are based on the frequency information alone, whereas the natural method makes use of two additional principles. One of these principles, based on stochastic resonance, is especially spectacular as it uses noisy elements of the sensory system in order to amplify weak auditory input signals. The design of the next generation of cochlear implants should therefore include noisy elements in a concept similar to that shown by nature.

Not surprisingly, the Section "Material & Structure" by comparison is more extensive than "Form & Construction" and "Information & Dynamics". This reflects a big focus on material science and related fields. But nevertheless one can find true international reputation and competence also in the other areas cited here.

References

1. A. El Albani et al., Large colonial organisms with coordinated growth in oxygenated environments 2.1 Gyr ago. Nature **466**, 100–104 (2010)
2. Ph.C.J. Donoghue, J.B. Antcliffe, Origins of multicellularity. Nature **466**, 41–42 (2010)
3. P. Ball, Life's lessons in design. *Nature* **409**, 413–416 (2001)
4. K. Rick, Economy and bionics, http://pgamb.up.edu.br/arquivos/pgamb/File/SummerSchool/Rick%20-%20Economy%20and%20Bionics%20.pdf [3/16/10]
5. E. Jorna (contr. ed.), Sustainable innovation. *The Organisational, Human and Knowledge Dimension* (Greenleaf Publishing Co., 2006) Sheffield, UK
6. N. Jursic, http://sammelpunkt.philo.at:8080/797/2/wimo/node4.html [07/08/10 21:31:50]
7. bmvit (Bundesministerium f. Verkehr, Innovation u. Technologie). BionIQA – Bionik Innovation und Qualifikation Austria. Wien (2010)
8. DTI (Department of Trade and Industry). Biomimetics: strategies for product design inspired by nature – a mission to the Netherlands and Germany. Report of a DTI Global Watch Mission (January 2007)
9. J. Matovic, Z. Jakšić, Bionic (nano)membranes. in *Biomimetics – Materials, Structures and Processes. Examples, Ideas and Case Studies*, ed. by P. Gruber et al., Series: Biological and Medical Physics, Biomedical Engineering (Springer Publishing, Series Editor Claus Ascheron, 2011)
10. I.C. Gebeshuber, et al. Biomimetics in tribology. in *Biomimetics – Materials, Structures and Processes, Examples, Ideas and Case Studies*, ed. by P. Gruber et al., Series: Biological and Medical Physics, Biomedical Engineering (Springer Publishing, Series Editor Claus Ascheron, 2011)
11. H.A. Abdel-Aal, M. El Mansori, Reptilian skin as a biomimetic analogue for design of deterministic tribo-surfaces. in *Biomimetics Materials, Structures and Processes, Examples,*

Ideas and Case Studies, ed. by P. Gruber et al., Series: Biological and Medical and Medical Physics, Biomedical Engineering, (Springer Publishing, Series Editor Claus Ascheron, 2011)
12. C. Hellmich, A. Fritsch, L. Dormieux, Multiscale homogenization theory: An analysis tool for revealing mechanical design principles in bone and bone replacement materials. in *Biomimetics – Materials, Structures and Processes. Examples, Ideas and Case Studies*, ed. by P. Gruber et al., Series: Biological and Medical Physics, Biomedical Engineering (Springer Publishing, Series Editor Claus Ascheron, 2011)
13. J. Stampfl, H.E. Pettermann, R. Liska, Bio-inspired cellular structures: Additive manufacturing and mechanical properties. in *Biomimetics – Materials, Structures and Processes. Examples, Ideas and Case Studies*, ed. by P. Gruber et al., Series: Biological and Medical Physics, Biomedical Engineering (Springer Publishing, Series Editor Claus Ascheron, 2011)
14. P. Gruber, Biomimetics in architecture [Architekturbionik]. in *Biomimetics – Materials, Structures and Processes. Examples, Ideas and Case Studies*, ed. by P. Gruber et al., Series: Biological and Medical Physics, Biomedical Engineering (Springer Publishing, Series Editor Claus Ascheron, 2011)
15. D. Kuhlmann, Biomorphism in architecture – speculations on growth and form. in *Biomimetics – Materials, Structures and Processes. Examples, Ideas and Case Studies*, ed. by P. Gruber et al., Series: Biological and Medical Physics, Biomedical Engineering (Springer Publishing, Series Editor Claus Ascheron, 2011)
16. W.E. Lorenz, Fractal geometry of architecture – fractal dimension as a connection between fractal geometry and architecture. in *Biomimetics – Materials, Structures and Processes. Examples, Ideas and Case Studies*, ed. by P. Gruber et al., Series: Biological and Medical Physics, Biomedical Engineering (Springer Publishing, Series Editor Claus Ascheron, 2011)
17. D. Bruckner, D. Dietrich, G. Zucker, B. Müller, Biomimetics in intelligent sensor and actuator automation systems. in *Biomimetics – Materials, Structures and Processes. Examples, Ideas and Case Studies*, ed. by P. Gruber et al., Series: Biological and Medical Physics, Biomedical Engineering (Springer Publishing, Series Editor Claus Ascheron, 2011)
18. M. Gföhler, Technical rebuilding of movement function using functional electrical stimulation. in *Biomimetics – Materials, Structures and Processes. Examples, Ideas and Case Studies*, ed. by P. Gruber et al., Series: Biological and Medical Physics, Biomedical Engineering (Springer Publishing, Series Editor Claus Ascheron 2011)
19. F. Rattay, Improving hearing performance utilizing natural auditory coding strategies. in *Biomimetics – Materials, Structures and Processes. Examples, Ideas and Case Studies*, ed. by P. Gruber et al., Series: Biological and Medical Physics, Biomedical Engineering (Springer Publishing, Series Editor Claus Ascheron 2011)

Part I
Material Structure

Chapter 2
Bionic (Nano) Membranes

Jovan Matovic and Zoran Jakšić

Abstract The goal of this chapter is to offer a concise and clear picture of the most important artificial nanomembrane-related procedures and technologies, including those for fabrication and functionalization, and to present the main properties and potential applications, stressing recent results in the field contributed by the authors. Nanomembranes are probably the most ubiquitous building block in biology and at the same time one of the most primordial ones. Every living cell, from bacteria to the cells in human bodies, has nanomembranes acting as interfaces between the cytoplasm and its surroundings. All metabolic processes proceed through nanomembranes and involve their active participation. Functionally, the man-made nanomembrane strives to mimic this most basic biological unit. The existence of the life itself is a proof that such a fundamental task can be performed. When designing artificial nanomembranes, the whole wealth of structures and processes already enabling and supporting life is at our disposal to recreate, tailor, fine-tune, and utilize them. In some cases, the obstacles are formidable, but then the potential rewards are stunning.

There is an additional advantage in bionic approach to nanomembranes: we do not have to use only the limited toolbox of materials and processes found in nature. Instead we are free to experiment with enhancements not readily met in natural structures – for instance, we may utilize nanoparticles of isotopes emitting ionizing radiation, even at lethal doses. We can introduce additional structures to our bionic nanomembranes, each carrying its own functionality, for instance nanoparticles or layers with plasmonic properties (e.g., to be used in sensing applications), target-specific binding agents (to improve selectivity) and carbon-nanotube support (to enhance mechanical strength). In this way, we are able to create meta-nanomembranes with properties exceeding the known ones (Jakšić and

J. Matovic (✉)
Institute for Sensor and Actuator Systems, Vienna University of Technology, Vienna, Austria
e-mail: Jovan.Matovic@tuwien.ac.at

Matovic, Materials 3:165–200, 2010). In this chapter, we present some small steps toward that goal.

2.1 Artificial Nanomembranes

Artificial nanomembranes are a very recent concept in micro and nanotechnologies [1, 2]. They may be defined as free-standing (self-supported) planar structures ranging 5–100 nm in thickness. A thickness of 5 nm corresponds to 15–20 atomic layers of silicon, which approaches the fundamental limits. Because of their unique structure, man-made nanomembranes with areas of several cm^2 and giant aspect ratios exceeding 1,000,000 are sufficiently robust to withstand laboratory handling without any special equipment and with only a modest degree of precaution [1,3,4]. Artificial nanomembranes simultaneously belong to two worlds: to that of the nanoelectromechanical systems (NEMS) because of their thickness and to that of the microelectromechanical systems (MEMS) owing to their area. They represent an artificial counterpart of the living cell membranes that divide the cytoplasm of the living cell from its environment and at the same time provide communication and enable their active interaction [3].

The most basic classification of man-made nanomembranes is inorganic and organic (macromolecular) ones. Early inorganic nanomembranes were in the form of homogenous (Cr, Pt) metallic films. Their areas were limited to about 1 μm^2 at a 6-nm thickness. These simple structures led to the next generation – the metal-composite nanomembranes. It is well known that composites can be tailored to have superior mechanical and thermal properties when compared with pure materials [5].

In a novel manufacturing process developed at Vienna University of Technology, metallic nanomembranes are modified by incorporating oxide and nitride nanoparticles within a metallic matrix [3]. The process is based on the standard MEMS technologies and does not require excessively expensive equipment and approaches. This technology combines reactive ion sputtering with simultaneous ion implantation into the substrate [3]. In this manner, nanoparticles are generated in situ during the deposition process itself. The result is a considerable enhancement of the mechanical properties enabling the fabrication of nanomembranes with areas 10^7-fold larger than previously possible (Fig. 2.1).

The second class includes organic (macromolecular) nanomembranes. This class is considerably greater than the inorganic structures, as there exists a plethora of organic molecules from which such nanomembranes may be assembled (see Fig. 2.3). There is also a wider choice of manufacturing processes available for organic nanomembranes compared with the inorganic ones.

During the 1930s, the Langmuir-Blodgett (LB) process was invented [6, 7]. LB nanomembranes are highly ordered. Being double-layered, they resemble biological membranes. Therefore, they were the first bionic membranes ever produced. Unfortunately, LB nanomembranes are very unstable and until now have found only niche applications [8].

2 Bionic (Nano) Membranes

Fig. 2.1 *Left.* SEM image of a metal-composite nanomembrane with lateral dimensions of 1.5 × 3.5 mm². Full thickness of the nanomembrane is only 7 nm. *Right* A photo illustrating the mechanical robustness of the metal-composite nanomembranes: the dimensions are identical to those in the image on the *left*. Current research in TU Wien – ISAS

More recent production methods include spin coating or thermal deposition of macromolecules based on thiol or silane compounds onto a substrate. However, unlike the LB process, these methods do not allow the molecular order in the films to be controlled; therefore, those films are hardly appropriate for bionic structures [9].

Finally, there is a relatively recently introduced method for film self-assembly that makes use of the alternating adsorption of oppositely charged macromolecules (polymers, nanoparticles, and proteins) – the layer-by-layer (LbL) technique [10]. The assembly of alternating layers of oppositely charged long-chained molecules is a simple process, which closely mimics the natural self-organization in living cells. It provides the means to form 5–500 nm thick films with monolayers of various substances growing in a preset sequence on any substrate. These nanomembranes have lower molecular order than LB films, but they have the advantage of high strength and easy preparation [11].

The properties of inorganic and organic nanomembranes are fundamentally different, the only common trait being miniscule thickness and large aspect ratio. Inorganic nanomembranes are mechanically robust and stable in harsh environments, but largely lack functionalization possibilities. However, polymer nanomembranes provide a variety of functional advantages over inorganic nanomembranes because they can have a wide range of source materials, tunable surface, and structure functionalities.

The practical utilization of nanomembranes is still in its infancy; however, some application fields seem rather promising. Some examples include new generations of photonic and chemical sensors, fluid separation, and energy conversion. The function of most of them resembles the biological cell nanomembranes. The essential step toward wider applications of artificial nanomembranes is the combination of inorganic and organic materials into a single structure. This may be done by applying laminated inorganic and organic layers and by incorporating various functional groups.

2.2 Biological Nanomembranes

Biological nanomembranes consist of lipid bilayers incorporating proteins and carbohydrates and typically have a very complex internal structure. The biological nanomembranes are highly functionally ordered, not simply geometrically ordered, Fig. 2.2. Science still has a way to go toward full understanding of the structure and function of the cell membranes, even in the simplest cases. However, the phenomenological reactions of a cell membrane to stimuli are better known: the membrane actively alters its permeability to respond to the specific molecule concentrations inside or outside the cell. It can transport molecules symmetrically or asymmetrically, passively by diffusion or actively by various "pumps". The key to this complex behavior is the membrane functionalization, which appeared at the very beginning of life itself.

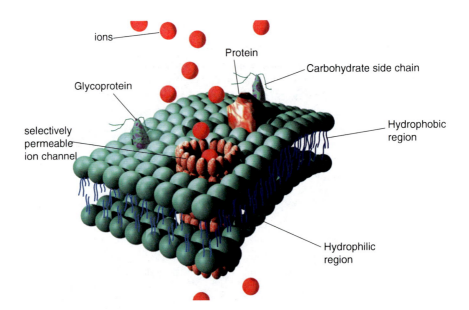

Fig. 2.2 Simplified representation of a lipid bilayer biological nanomembrane

2.3 Functionalization of Artificial Nanomembranes Toward Bionic Structures at ISAS: TU Wien

Bionic or biomimetic science may be defined as the study of biological systems as models for design and engineering of materials and systems. The model for the artificial nanomembranes, at least regarding their form and dimensions, are obviously the biological cell membranes. There is one fundamentally different trait of the artificial nanomembranes at their current state of development – their striking lack of complex functionalities. This is in stark contrast with the biological structures where such functionalities are not only common but also at the core of their function. This may appear like a very serious drawback of the man-made structures. However, one has to bear in mind that the research field of the artificial nanomembranes was conceived very recently, and the main body of literature being published within the last several years. Imparting complex biomimetic properties to nanomembranes is even more recent and is literally at its onset. Witnessing the present accelerating progress toward more complex functional structures, rapid advances may well be expected in the near future. The current research at the ISAS of nanomembranes with bionically enhanced functionality is directed exactly toward this goal [33]. Currently, it is proceeding along two main and seemingly disparate lines, fuel cell-based energy harvesting and detection of long wavelength infrared (IR) radiation.

2.3.1 Nanomembrane-Based Bionic Structures for Energy Harvesting

Stable and continuous energy supply is one of the very foundations of a peaceful and prosperous society. Considered from the thermodynamical point of view, the forms of energy currently used can be basically classified into two main groups: the low quality thermal energy, and the high quality energy like electricity or mechanical work.

Low quality energy (i.e., heat) is still mostly produced by burning accumulated fossil fuels. Other alternatives for obtaining thermal energy in large quantities and at high temperatures still fail from various causes. For instance, solar energy is highly dispersed (max. $2\,\text{kW}/\text{m}^2$, rapidly declining at higher latitudes). Also, it is very difficult and technologically challenging to obtain high amounts of such energy in industrial plants and to further deliver it. Other energy sources such as biomass are helpful, but have many drawbacks and definitely do not suffice for the increasing energy demands of the modern society.

The situation with the production of electric power is even more complex. Nowadays the basic principles in power production remain unchanged when compared with those at the beginning of the industrial era. An energy carrier, typically fossil fuel, is used to generate heat, which is further transferred to a fluid.

In the next stage, the energy of the heated fluid is converted first to mechanical work and then to electricity in the plant's generators. The overall efficiency of this process is fundamentally limited by the Carnot cycle to 30–50%. Generally, the conversion of heat into the mechanical work is a low efficiency process. Unused heat is released into the environment, causing heat pollution, together with the emission of carbon dioxide and other harmful pollutants. Even a nuclear power plant deviates little from this scheme, indeed emitting no carbon dioxide, but churlishly leaving the problem of radioactive waste to the next generations.

Another process to convert chemical energy bound in fuel molecules to electricity is the electrochemical conversion. Its advantage is that, being an isothermal process, it is free of the Carnot cycle limitation. Examples of man-made electrochemical generators are batteries and fuel cells (FC). Although they share many similarities, FC substantially differ from batteries. They do not internally contain fuel and continue to convert chemical energy into electric energy (and some heat) as long as fuel and oxidant are supplied from external sources.

The operation principle of a proton exchange membrane (PEM) fuel cell is shown in Fig. 2.3. The central part of PEM FCs is a thin membrane, nonconductive but highly permeable to hydrogen ions (protons). On the anode side, the fuel, in this case hydrogen gas (H_2), is dissociated to protons and electrons by means of nanodispersed platinum catalyst. Electrons are collected on the anode side and made to flow into an external circuit, generating electricity. Simultaneously, protons migrate through the membrane toward the cathode. On the cathode side, the protons and electrons recombine with oxygen from the atmosphere, again with the assistance of the platinum catalyst. The basic reaction chain is given below:

$$H_2 \rightarrow 2H^+ + 2e^- \ (E_r = 0\,V)$$

$$O_2 + 4H^+ + 4e^- \rightarrow 2H_2O \ (E_r = 1.23\,V)$$

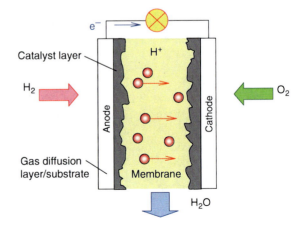

Fig. 2.3 Schematic of electrochemical reactions in proton exchange membrane fuel cell (PEM FC). Among many types of fuel cells existing today, the PEM FCs have the highest specific energy density and work at a temperature of about 100 °C

2 Bionic (Nano) Membranes

The final process of hydrogen oxidation is water. The FC are essentially zero emission generators. The practical efficiency of the conversion process is still around 40%, although the theoretical limit is close to 90% [12].

The PEM or proton exchange barrier is the critical part of a PEM fuel cell. The basic function of the membrane is to enable proton transport, while being simultaneously impermeable for electrons and gases. Typically, membranes for the PEM, FC are made up of perfluorocarbon-sulfonic acid (PSA) ionomer. The best-known material of this class is Nafion (Trademark of DuPont). Nafion has a unique interpenetrating structure of hydrophobic perfluorocarbon regions (providing thermal and chemical resistance, mechanical strength, and gas diffusional resistance) combined with hydrophilic regions of water clusters surrounding charged sulfonic acid groups (which allow selective proton transport), Fig. 2.4. For these reasons, Nafion, although already introduced in the mid-1960s, is still considered the benchmark against which most of the new materials are compared [13, 14].

However, Nafion and Nafion-like membranes suffer from several deficiencies that limit the efficiency of FCs to about 50% of the theoretical limit, but also make the production of FCs complex and expensive.

Ideally, a linear path between the two electrodes would provide the most efficient proton transport. However, the proton transport in Nafion polymer follows a random path, because of spatially disordered pores and the pore-geometry characterized by broad size distributions. The next and fundamental deficiency of Nafion materials is the prevalent mechanism of proton transport through the polymer nanochannels via the "vehicle" mechanism. At the molecular level, proton transport may follow two principal mechanisms:

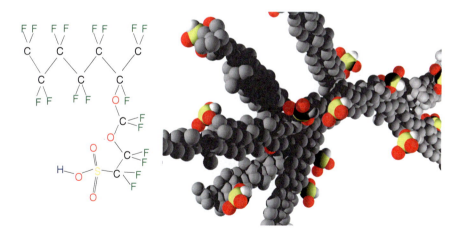

Fig. 2.4 Chemical and physical structure of Nafion. *Grey spheres* represent the hydrophobic backbone structure (Teflon) and *red* are the hydrophilic sulfonated clusters connected by an oxygen ion to the backbone chain

(a) "Vehicle" mechanism, i.e., diffusion of protons bound to carriers like $H^+(H_2O)_n$ ions, which travel along a nanochannel under electro-osmotic drag force, from the anode to the cathode side. Therefore, the vehicle proton transport is unavoidably associated with water molecule transport, resulting in water flow through the membrane [15].

(b) Proton hopping mechanism (Grotthuss transport). In some confined geometries, water molecules are aligned along a chain (called the "water wire"). If an excess proton is brought to the chain, it forms a protonic "defect." This defect moves by diffusion along the hydrogen bond chain of water molecules by alternately forming covalent bond with a molecule, which then releases another proton by splitting its covalent bond. Thus, the proton "hops" from one water molecule to the next along a quasi one-dimensional water wire [16]. The same effect may also occur in other liquids with hydrogen bonds. The water molecules within the water wire are immobile and transport of protons involves only temporary modification of covalent bonds within the water molecule during proton hopping [17,18]. Instead of H^+ $(H_2O)_n$ molecules, which exist in Nafion channels, water wires almost entirely conduct H^+ ions, i.e., protons. Hence, in the Grotthuss transport, there is no net water flow within the channel, Fig. 2.5. The path in this case is much closer to a straight line.

Now we can correlate the proton transport of artificial membranes such as Nafion with the biological nanomembranes. The flow of protons through a biological cell membrane is as fundamental to the cell metabolism as the flow of water, being coupled to the energy cycle that fuels the cell metabolism. It is advantageous for a cell to have separate transport of water (to maintain the internal pressure) and protons (necessary for cell energetics). The water balance within a cell is maintained

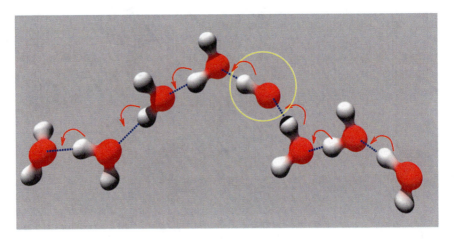

Fig. 2.5 Proton hopping mechanism confined to a one-dimensional water wire, the dominant proton transport mechanism in biological and bionic structures

by aquaporin water channels (Nobel Price in Chemistry 2003; [19]). Aquaporins are proteins with a hole along their middle with somewhat paradoxical behavior: they allow the flow of water molecules but are impermeable to protons that are much smaller than water molecules. In contrast, proton nanochannels are highly selective to protons (i.e., hydrogen ions) but are impermeable to water molecules. Any significant imbalance in proton or water concentration immediately causes cell dysfunction or death. Because of that both classes of channels are actively gated by the cell control mechanisms enabling control of proton concentration and internal osmotic pressure inside the cytoplasm within narrow margins. These molecular assemblies were definitely the first proton transport structures designed by nature and can be found in all organisms. A comprehensive analysis of proton channels is given by [20].

Ungated proton channel structures also exist in the natural world, although are less common. The most extensively studied one is gramicidin A (gA). gA is a small, double helix shaped molecule with hydrophilic body and hydrophobic ends (Fig. 2.6, left).

Each of the helices contains a water-filled channel, a natural water wire. Gramicidin channels have astonishingly high proton conductivity, reaching 2×10^9 H$^+$/s and thus surpassing any other known ion channel. The primary role of gA in natural environment is the self-defense weapon of *Bacillus brevis*. gA can disable the alien cell ability to maintain proton concentration in the cytoplasm, causing death of the invader cell.

As an organic molecule gA is relatively stable against weak acids and at moderately elevated temperatures. Further gA can be readily incorporated into artificial phospholipid bilayer membranes. These properties of gA already attracted attention of the industry with respect to proton conductive structures (Hewlett

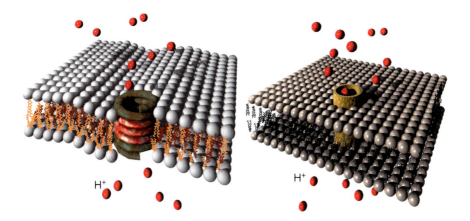

Fig. 2.6 (*Left*) gA helix molecule embedded in the biological phospholipids bilayer membrane. (*Right*) Bionic proton conductive structure. The artificial lipid-bilayer includes an array of proton conductive nanochannels, for example gA [21]

Packard) (Jackson and Jeon 2004). However, the stability and durability of the gA protein molecule is still insufficient for direct application in FCs at temperatures around 100 °C for prolonged periods of time. An idea is therefore to use alternative mechanisms to build artificial ion channels (Fig. 2.6, right).

Water-filled carbon nanotubes (CNTs) appear also as the candidate for a new generation of proton conductive structures. In this case, CNTs play the role of the ion channel within the living cell. CNTs are mechanically robust, chemically and thermally exceptionally stable, and have an integrated nanochannel. CNTs normally have hydrophobic walls, which can be made hydrophilic and their channel filled with a one-dimensional water wire theoretically resembling a gA structure [22], Fig. 2.7. A similar structure utilizing calixarene molecules with well-defined channels was also proposed [23].

The general direction in the further development of proton conductive membranes from the stochastic structure of Nafion-like materials toward highly efficient bionic nanomembranes seems to be marked out. However, the technological difficulties on the way toward practical realizations are tremendous. Many groups around

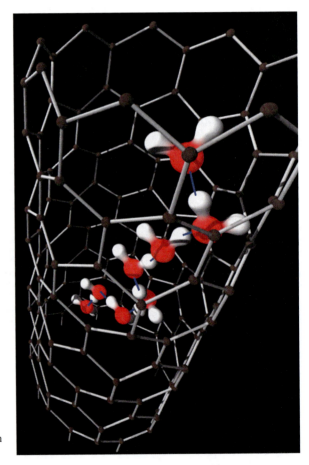

Fig. 2.7 CNT filled with water to form proton conductive water wire. The proton conductive mechanism is Grotthuss transport

the world are engaged in development of bionic proton conductive membranes. The TU WIEN, ISAS joined that circle through an EU FP7 project MultiPlat, "Biomimetic Ultrathin Structures as a Multipurpose Platform for Nanotechnology-Based Products." As the project is at a relatively early stage, not all details are currently available for publication.

2.3.2 Nanomembranes as Bionic Detectors of Electromagnetic Radiation

In the course of evolution, starting practically from the very emergence of life, most living organisms developed sensitivity to some wavelength range of electromagnetic radiation. Organisms like plants are able to perceive the direction of the electromagnetic radiation in the range 0.4–0.7 μm, since it ensures the energy necessary for their development (heliotropism).

In the case of organisms able for active movements (animals), the receivers of electromagnetic radiation evolved over time into their sense of vision. The information obtained through it helps the animals to orient, communicate, find food, and to escape from enemies.

Anthropocentrically, man called his electromagnetic sense the sense of vision, and the narrow electromagnetic range from 0.37 to 0.75 μm to which his/her sense is sensitive became the "visible light." However, some animals developed senses for other electromagnetic ranges, which cannot be registered by man's unaided eye. The wavelengths of maximum sensitivity of these senses suit the specific needs of these species, i.e., enable them to obtain the maximum of the information important for their survival. For instance, the common bee (*Apis mellifera*) sees radiation from yellow light to UV-A range (300 nm). Clouds are transparent at a wavelength of 365 nm, which ensures a safe solar navigation to bees and at the same time facilitates them to find food, since flowers with ultraviolet reflecting pigments clearly stand out against the background of green leaves. For humans, these flowers are simply white.

Other species developed senses for the IR range. IR radiation carries a significant portion of the information about our environment. Actually, more than 50% of the solar radiation (blackbody at 6,000 K) belongs to the IR range. A body heated to a temperature of 1,000 K (red heat) emits only 0.0002% of its energy in the visible range, while even 50% of the energy (and thus information) belongs to a wavelength range above 4.2 μm. Our environment at a temperature of 300 K emits solely in the IR. It is not surprising that the past few decades have seen a tremendous interest and also an unsurpassed growth in the technology of IR detection.

Two intensively studied species that developed specialized sensory organs for IR radiation are the insect *Melanophila acuminata* (jewel beetle) and the snakes from *Crotalidae* and *Boidae* families (Fig. 2.8a).

Jewel beetles, which lay their eggs into burned pinewood in recently burned forests, have organs sensitive to IR radiation, with a maximum sensitivity at

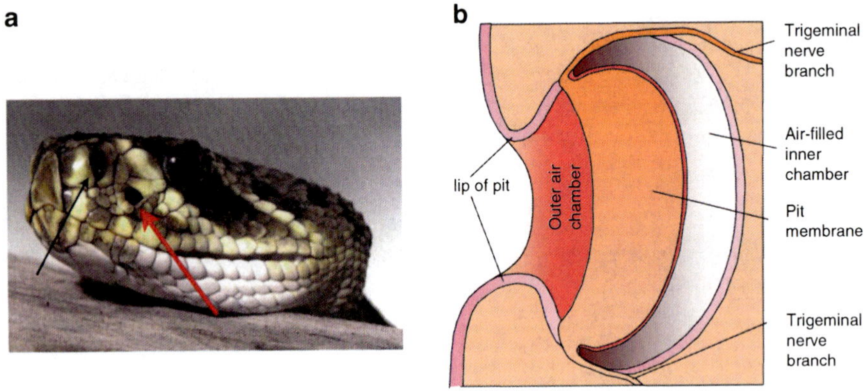

Fig. 2.8 (**a**) Rattlesnake, genus *Crotalus*. The *arrow* on the *right* points to the pit organ, the one on the *left* (*black*) shows the nostril (*Source*: Wikimedia Commons, under GNU Free Documentation License); (**b**) Cross-section of the pit organ where a membrane at the bottom of the pit serving as thermal infrared detector (denoted as Pit membrane) is clearly visible

2.8–3.5 μm, corresponding to the signature of a forest fire [24]. The IR systems of the snakes belonging to the subfamily *Crotalidae* and to the family *Boidae* are especially complex. IR organs are located in the head of the snake as a linear array of tiny apertures (pit organs) [25], Fig. 2.8a). Pits actually represent a pinhole optic system (Camera obscura) as no organic materials are transparent to 10 μm IR radiation (Fig. 2.8b). This wavelength corresponds to the maximum of the Planck blackbody radiation spectrum at a temperature close to 40 °C, which is the radiation of warm-blooded animals representing the basic pray of these snakes [26, 27].

It is accepted in biophysics that IR sensing in snakes is a thermal process [27]. IR photons are absorbed in the membrane via molecular resonant frequencies inherent to the chemical structure of the tissue. In essence, this energy transfer from the IR photons causes the molecules within the system to "vibrate" on the molecular level [28].

Prior to the development of the bionic nanomembrane-based IR sensors, one would do well to review the state of artificial IR detectors. Basically, IR detectors fall into two categories: photon and thermal detectors. The photon detectors of long wavelength IR radiation are made from semiconductor materials, and contemporary detectors must be cooled to achieve high sensitivity. Because of their requirements for cryogenic cooling, the photonic IR detectors tend to be expensive, complex, bulky, and difficult to maintain [29].

Man-made thermal detectors are based on the excitation of phonons or electrons in a solid by incident IR photons. This occurs through a cascade of different physical processes, and thus the incident IR energy is converted into random motion of lattice ions, i.e., into heat. This process is similar to those in biological sensors.

The sensitivity of modern thermal detectors approaches that of the semiconductor ones, and at the same time there is no need for cryogenic cooling. However, the response time of thermal detectors (including biological sensors) is radically inferior compared with that of the photon detectors. Thermal detectors respond relatively slowly (of the order of 10^{-3} s) compared with the photon detectors (of the order of 10^{-8} s).

A thermal detector response may be essentially characterized by two figures of merit: its specific detectivity $D*$ and response time τ as:

$$D* = \sqrt{\frac{A}{4k\varepsilon T^2 G}},$$

$$\tau = \frac{c}{G},$$

where A is the detector area, G is its thermal conductivity toward the ambient, ε is the detector emissivity, T is the detector temperature, and c is the thermal mass of the detector. To reach a high specific detectivity, the thermal conductivity of an IR detector should be the lowest possible to prevent leakage of heat. The sensitive IR organ in *Python molurus bivittatus* is therefore a freestanding membrane surrounded by air, thus minimizing thermal losses. Low thermal capacity is essential for a high response speed of a detector. The python sensitive membrane is only 15-μm thick, including blood vessels and thermosensitive nerve ends. Further, nerves in the membrane are very sensitive to small temperature variations. It is interesting that the sensitivity of the thermal receptors in the python's IR organ is not higher than that of the corresponding receptors in the human skin. However, our thermal receptors are located relatively deep under the skin surface. This naturally designed IR detector is relative sensitive and comparably fast for a thermal detector.

Bionic IR detectors attract increasing R&D interest nowadays [28, 30, 31]. However, there is an important difference in the development of bionic IR detectors compared with the development of the previously elaborated bionic proton-conductive structures. The performance of the artificial proton conductive structures still lags behind their natural counterparts, while the performance of the contemporary man-made IR detectors exceeds the natural ones. The principal rationale for this situation is that our detector technology exploits a wider range of materials than are available in biological structures. Some of these materials are highly toxic (Mercury Cadmium Telluride, MCT for example), others require excessively high or cryogenic temperatures during production processes, high pressure, etc. We conclude that the application of bionic principles, not materials themselves, still can result in vastly improved performance of IR detectors.

Bionic nanomembranes are a logical further step in developing thermal IR detectors. They can be manufactured to be highly sensitive to small temperature variations. At the same time, their thermal mass is generally very close to or just about equal to the fundamental limits. For example, graphene, a monatomic planar structure of carbon atoms arranged in a benzene ring network, is still sufficiently robust to be incorporated in a structure similar to the snake pinhole-based IR sensing organ. It is hard to imagine a structure with a lower thermal mass.

TU Wien actively works on bionic IR detectors within an FWF (Austrian Science Fund) project. Currently, the thickness of the nanomembranes used in the

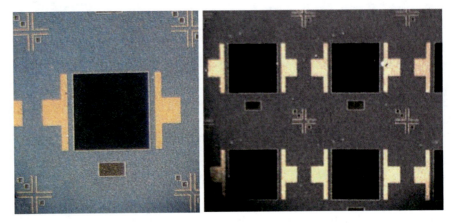

Fig. 2.9 Microphotograph of the experimental bionics IR detector based on nanomembrane. Current research at TU Wien – ISAS within a FWF Project

experimental detector is 6–7 nm and will be reduced even more in the future. The time constant of this thermal detector should be in the nanoseconds range, compared with milliseconds of the contemporary thermal devices (Fig. 2.9).

2.4 Conclusion

In this chapter, we have reviewed the most important procedures, technologies, and structures related to bionic nanomembranes as the artificial counterparts to the biological lipid bilayer membranes. We stressed our related contributions and presented our current works. Among these, we dedicated special attention to two particular biomimetic applications of the artificial nanomembranes: the proton-exchange membranes for the novel generation of FC and the nanomembrane-based thermal detectors mimicking the function of snake pit organs for IR vision.

When designing an artificial nanomembrane, one does not have to use the severely limited toolbox of nature. The available materials, structures, and processes can be far extended both into the inorganic and organic world. We may use various procedures and conditions usually not encountered with the living tissue, including high temperatures, aggressive media, radiation, etc. The much higher degree of design freedom and possibilities to include properties not found in natural materials already helped our research related to thermal detectors, resulting in performance exceeding that found in the living beings.

The research of artificial nanomembranes has the following three main goals:

- To build artificial nanomembranes using an extended toolbox
- To ensure membrane functionalization; for instance, to enable artificial transmembrane passageways (various engineered nanopores and artificial ion

channels), to implement structural reinforcements and many other additional functionalities
- To find ways for the most widespread application of the new building blocks in different practical applications and to optimize them

Different fields of application include nanoelectrochemistry and nanocatalysis, life science/biomedicine, optical engineering, photonics, and plasmonics, the use in various sensors (Jiang et al. 2004) including mechanical, thermal, chemical, biological ones, etc. Of large importance are protection (increasing mechanical stabilization and durability, avoiding tribological problems) and separation (nanofiltering, selective ion transfer) [32]. The bionic properties of nanomembranes enable us to rethink and reinvent some classical applications of ultrathin structures, this time by imparting them quasi-living features. Examples include transmembrane transport in proton-exchange FC, enhancement of various photochemical processes including energy conversion and hydrogen generation, artificial photosynthesis, to mention just a few. Bionic nanomembranes have a potential to improve environmental protection, to bring breakthroughs in life science, to enable the production of clean energy, and to contribute in numerous other ways to an enrichment of the overall quality of life.

Acknowledgements This work was funded by the Austrian Science Fund (FWF) within the project L521 "Metalcomposite Nanomembranes for Advanced Infrared Photonics" and within EU FP7 Framework project "MultiPlat" (Biomimetic Ultrathin Structures as a Multipurpose Platform for Nanotechnology-Based Products).

References

1. C. Jiang, S. Markutsya, Y. Pikus, V.V. Tsukruk, Freely suspended nanocomposite membranes as highly sensitive sensors. Nat. Mat. **3**, 721–728 (2004)
2. J. Matovic, Z. Jakšić, in *Micro Electronic and Mechanical Systems*, ed. by K. Takahata. Nanomembrane: A New MEMS/NEMS Building Block (In-Tech, Vukovar, 2010), pp. 61–84
3. J. Matovic, Z. Jakšić, Simple and reliable technology for manufacturing metal-composite nanomembranes with giant aspect ratio. Microel. Engg. **86**, 906–909 (2009)
4. R. Vendamme, S.Y. Onoue, A. Nakao, T. Kunitake, Robust free-standing nanomembranes of organic/inorganic interpenetrating networks. Nat. Mat. **5** (6) 494–501 (2006)
5. K. Friedrich, S. Fakirov, Z. Zhang, *Polymer Composites: From Nano- to Macro-Scale* (Springer, Berlin 2005)
6. K.B. Blodgett, I. Langmuir, Built-up films of barium stearate and their optical properties. Phys. Rev. **51**, 964–982 (1937)
7. M.C. Petty, *Langmuir-Blodgett Films: An Introduction* (Cambridge University Press, Cambridge, UK, 1996)
8. H. Endo, M. Mitsuishi, T. Miyashita, Free-standing ultrathin films with universal thickness from nanometer to micrometer by polymer nanosheet assembly. J. Mat. Chem. **18** (12), 1269–1400 (2008)
9. H. Watanabe, T. Kunitake, Freestanding, 20 nm thick nanomembrane based on an epoxy resin. Adv. Mater. **19**, 909–912 (2007)

10. Y. Lvov, I. Ichinose, T. Kunitake, Assembly of multicomponent protein films by means of electrostatic layer-by-layer adsorption. J. Am. Chem. Soc. **117**, 6117–6122 (1995)
11. Z. Liang, K.L. Dzienis, J. Xu, Q. Wang, Covalent layer-by-layer assembly of conjugated polymers and cdse nanoparticles: multilayer structure and photovoltaic properties. Adv. Funct. Mater. **16** 4, 542–548 (2006)
12. G. Hoogers, *Fuel cell Technology Handbook* (CRC Press, New York, 2003)
13. S.J. Hamrock, M.A. Yandrasits, Review: proton exchange membranes for fuel cell applications. J. Macromol. Sci. C Polym. Rev. **46** 219–244 (2006)
14. R.G. Rajendran, Polymer electrolyte membrane technology for fuel cells. MRS Bull. **30**, 587–590 (2005)
15. K.D. Kreuer, Proton conductivity: materials and applications. Chem. Mater. **8**, 610–641 (1996)
16. A. Noam, The Grotthuss mechanism. Chem. Phys. Lett. **244**, 456–462 (1995)
17. S.H. Chung, O.S. Andersen, V. Krishnamurthy (eds.), *Biological Membrane Ion Channels: Dynamics, Structure, and Applications* (Springer, Berlin, 2006)
18. C.J.T de Grotthuss, Sur la décomposition de l'eau et des corps qu'elle tient en dissolution à l'aide de l'électricité galvanique. Ann. Chim. **58**, 54–73 (1806)
19. B. Eisenberg, Why can't protons move through water channels. Biophys. J. **85**, 3427–3428 (2003)
20. T.E. Decoursey, Voltage-gated proton channels and other proton transfer pathways. Physiol. Rev **83**, 475–579 (2003)
21. W.B. Jackson, Y. Jeon, Highly discriminating, high throughput proton-exchange membrane for fuel cell applications. US Patent Appl. 0,191,599, 2004
22. D.J. Mann, D.J. Halls, Water alignment and proton conduction inside carbon nanotubes. Phys. Rev. Lett., **90** (19), 195–503 (2003)
23. L. Jae-jun, J. Myung-sup, K. Do-yun, L. Jin-gyu, M. Sang-kook, Solid acid, polymer electrolyte membrane including the same, and fuel cell using the polymer electrolyte membrane. US Patent Appl. 0,082,248, 2007
24. G.W. Evans, Infra-red receptors in *Melanophila acuminata*, Nature **202**, 211 (1964)
25. T.H. Bullock, R.B. Cowles, Physiology of an infrared receptor: the facial pit of pit vipers. Science **115**, 541–543 (1952)
26. M.S. Grace, D.R. Church, C.T. Kelley, W.F. Lynn, T.M. Cooper, The Python pit organ: imaging and immunocytochemical analysis of an extremely sensitive natural infrared detector. Biosens. Bioelectr. **14**, 53–59 (1999)
27. E.A. Newman, P.H. Hartline, The infrared "vision" of snakes. Sci. Am. **246** (3), 98–107 (1982)
28. V. Gorbunov, N. Fuchigami, M. Stone, M. Grace, V.V. Tsukruk, Biological thermal detection: micromechanical and microthermal properties of biological infrared receptors. Biomacromolecules **3**, 106–115 (2002)
29. A. Rogalski, Infrared detectors: status and trends. Prog. Quant. Electron. **27** 59–210 (2003)
30. R. Bogue, From bolometers to beetles: the development of thermal imaging sensors. Sensor Rev. **27** (4) 278–281 (2007)
31. O. Yavuz, M. Aldissi, Biomaterial-based infrared detection. Bioinsp. Biomim. **3**, 035007, (2008)
32. C.C. Striemer, T.R. Gaborski, J.L. McGrath, P.M. Fauchet, Charge- and size- based separation of macromolecules using ultrathin silicon membranes. Nature **445**, 749–753 (2007)
33. Z. Jakšić, J. Matovic, Functionalization of artificial freestanding composite nanomembranes. Materials **3** (1), 165–200 (2010)

Chapter 3
Biomimetics in Tribology

I.C. Gebeshuber, B.Y. Majlis, and H. Stachelberger

Abstract Science currently goes through a major change. Biology is evolving as new Leitwissenschaft, with more and more causation and natural laws being uncovered. The term 'technoscience' denotes the field where science and technology are inseparably interconnected, the trend goes from papers to patents, and the scientific 'search for truth' is increasingly replaced by search for applications with a potential economic value. Biomimetics, i.e. knowledge transfer from biology to technology, is a field that has the potential to drive major technical advances. The biomimetic approach might change the research landscape and the engineering culture dramatically, by the blending of disciplines. It might substantially support successful mastering of current tribological challenges: friction, adhesion, lubrication and wear in devices and systems from the meter to the nanometer scale. A highly successful method in biomimectics, the biomimicry innovation method, is applied in this chapter to identify nature's best practices regarding two key issues in tribology: maintenance of the physical integrity of a system, and permanent as well as temporary attachment. The best practices identified comprise highly diverse organisms and processes and are presented in a number of tables with detailed references.

I.C. Gebeshuber (✉)
Institute of Microengineering and Nanoelectronics (IMEN), Universiti Kebangsaan Malaysia, 43600 UKM, Bangi, Selangor, Malaysia
and
TU BIONIK Center of Excellence for Biomimetics, Vienna University of Technology, Getreidemarkt 9/134, 1060 Vienna, Austria
and
Institute of Applied Physics, Vienna University of Technology, Wiedner Hauptstrasse 8–10/134, 1040 Vienna, Austria
and
AC^{2}T Austrian Center of Competence for Tribology, Viktor Kaplan-Straße 2, 2700 Wiener Neustadt, Austria
e-mail: gebeshuber@iap.tuwien.ac.at

As next step, detailed investigations on the relevant properties of the best practices identified in this chapter shall be performed, and the underlying principles shall be extracted. Such principles shall then be incorporated into devices, systems and processes; and thereby yield biomimetic technology with increased tribological performance. To accelerate scientific and technological breakthroughs, we should aim at having a context of knowledge: the gap between scientific insights and technological realization should be bridged. To prevent being trapped in the inventor, innovator or investor gaps, a cross dialogue is necessary, a pipeline from 'know-why' to 'know-how' to 'know-what'. This is specifically of relevance in tribology, since tribological research is ultimately linked to real-world applications. Applying biomimetics to tribology could provide such a pipeline.

3.1 Introduction: Historical Background and Current Developments

Science currently goes through a major change: in biology, more and more causation and natural laws are being uncovered [1]. Biology has changed during the recent decades: it transformed from a rather descriptive field of research to a science that can – in terms of concepts, basic ideas and approaches – be acknowledged and understood by researchers coming from 'hard sciences' (such as physics, chemistry, engineering and materials science) including tribologists (Fig. 3.1) [2]. Tribology relies on experimental, empirical, quantifiable data or the scientific method, and focuses on accuracy and objectivity [3, 4]. The amount of causal laws in this new

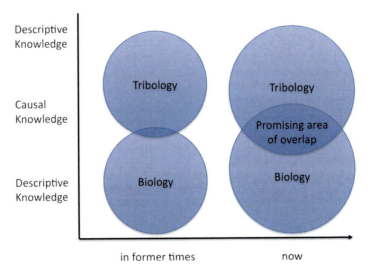

Fig. 3.1 The increasing amount of causal laws in biology generates promising areas of overlap with tribology

biology (indicated by the ratio of causal vs. descriptive knowledge) is steadily growing and a new field that can be called 'Biological Physics' is emerging [1]. The languages of the various fields of science increasingly get compatible, and the amount of collaborations and joint research projects between researchers coming from the 'hard sciences' and biologists have increased tremendously over the last years. Still, there is a large gap between the natural sciences and humanities [5].

The term 'technoscience' characterizes a field in which technology and science are inseparably interconnected. This characteristic hybrid form is, for instance, seen in the atomic force microscope – a symbol for both nanoscience and nanotechnology. This tool not only allows for basic scientific investigations, but also for manipulation and engineering at very small scales. In technoscience, there is no clear distinction between investigation and intervention. Even more, by investigation already interventions may be made. Application-oriented biomimetics can be denoted as 'technoscience'.

Traditionally, engineers are interested in what works, i.e. what functions and is useful, and are hence rather pragmatic, whereas scientists are interested in explanations, hypotheses and theories that reflect a rather different stance. For scientists, experiments are meant to try and prove or falsify a hypothesis or theory. The practical aspects of experiments, i.e. the potential applicability, do not belong to science but to technology. 'While traditional conceptions of science foreground the formulation and testing of theories and hypotheses, technoscience is characterized by a qualitative approach that aims to acquire new competencies of action and intervention' [6]. Of course, also pure scientific theories are a basis or prerequisite for technology, but it is not necessary to have an application in mind before a scientific investigation, which is a characteristic of the field of technical biology [7]. Living nature is seen from an engineering viewpoint, or even nature itself is thought of as an 'engineer' who is facing technical problems.

In biomimetics, materials, processes and systems in nature are analyzed; the underlying principles are extracted and subsequently applied to science and technology [7–10]. Biomimetics is a growing field that has the potential to drive major technical advances [1, 11, 12]. It might substantially support successful mastering of current tribological challenges. The biomimetic approach can result in innovative new technological constructions, processes and developments [7]. Biomimetics can aid tribologists to manage the specific requirements in systems or product design, to integrate new functions, to reduce production costs, to save energy, to cut material costs, to redefine and eliminate 'waste', to heighten existing product categories, to define new product categories and industries, to drive revenue and to build unique brands [13, 14].

Gebeshuber and Drack [7] distinguished two methods of biomimetics: biomimetics by analogy and biomimetics by induction, to which the different activities in the field can be assigned. Biomimetics by analogy starts with a problem from technology and tries to find analogous problems in nature with the respective solutions that might also be useful in the technology. Biomimetics by induction refers to ideas that stem from basic science approaches in biology, with no intention for applications as a motivation in the first place.

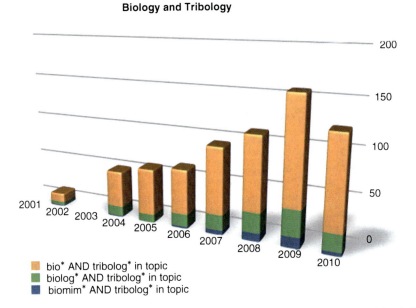

Fig. 3.2 The number of scientific publications in the years 2001–2010 with explicit relation between biology and tribology
Source: ISI Web of Knowledge, Thomson Reuters, Citation Databases: SCI-EXPANDED (2001-present), CPCI-S (2004-present). http://www.isiknowledge.com, (accessed 5 May 2010)

Biomimetics is yet another example for the increasing dissolution of disciplines that are found in science, together with the development of highly specialized domains. Interdisciplinary work with a specific focus (e.g. the functional design of interacting surfaces by means of nanotechnology) requires input from more than one classical discipline (in this example: physics, chemistry, biology, mechanical engineering, electronics and tribology). Recurrent concepts in biomimetics can easily be transferred to technology [1, 7, 8].

The amount of scientific papers that link biology to tribology is increasing (see Fig. 3.2). However, there is still a large unexplored body of knowledge that deals with lubrication and wear in biology but that has not yet been linked extensively to technology (Fig. 3.3).

3.2 Biology for Engineers

Engineers may not be primarily interested in evolution or taxonomy. Yet, basic knowledge about typical reactions of biological organisms or groups of organisms to conditions imposed by natural and human activities might prove beneficial for their work. Biology for engineers should be principle-based, viewed as a system

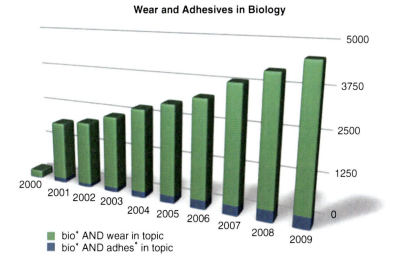

Fig. 3.3 The number of scientific publications in the years 2000–2009 dealing with either wear or adhesives in biology comprise a huge yet unexplored amount of inspiration for technology
Source: ISI Web of Knowledge, Thomson Reuters, Citation Databases: SCI-EXPANDED (2001-present), CPCI-S (2004-present). http://www.isiknowledge.com, (accessed 5 May 2010)

and might lead to predictive expectations about typical behavioural responses [15, Table 3.1].

Recurring principles of biology are correlation of form and function, modularity and incremental change, genetic basis, competition and selection, hierarchy and multi-functionality [16, 17].

General principles that can be applied by engineers who are not at all involved in biology have been distilled [18]. These basic principles comprise integration instead of additive construction, optimization of the whole instead of maximization of a single component feature, multi-functionality instead of mono-functionality, energy efficiency and development via trial-and-error processes. Systematic technology transfer from biology to engineering thereby becomes generally accessible.

Knowledge about the responses of biological systems may lead to useful products and processes, might increase the ability of engineers to transform information from familiar systems to unfamiliar ones and might help to avoid unintended consequences of emerging technologies.

Nachtigall promoted analogy search and states that the nature of qualitative analogy research is an impartial, open-minded comparison. He presents numerous examples of insect micromorphology and relates functional mechanisms to technological examples in a visual comparison [19].

In biomimicry, nature is seen as model and mentor (and measure for sustainability). Models in nature are studied, and forms, processes, systems and strategies are emulated to solve human problems – sustainably. Biomimicry is a new way of viewing and valuing nature. It introduces an era based not only on what we can

Table 3.1 Possible extrapolation of biological responses to technical systems

Biological responses [15]	Possible extrapolation to technical systems
Organisms die without water, nutrients, heat sources and sinks and the right amount of oxygen	Proper energy management
Organisms become ill in the presence of wastes	Proper waste management
Organisms modify their environments	Consider two way interaction of device with environment
Extra energy will be spent on adaptations	Rather adapt than completely change
Organisms, if possible, will move to friendlier environments	Choose promising niches
Organisms will evolve under environmental pressures	Reactive responsive adaptive devices
Crowding of organisms produces stress	Information management in an era of over-information
Organisms are affected by chemical and mechanical stresses	Reactive devices
Optimization is used to save energy and nutrient	Resourcefulness
Organisms alter themselves to protect against harsh environments	Adaptive devices
Organisms cooperate with other organisms	Sharing of data and results with other devices in the same system
Organisms compete with other organisms	Input from other devices is used to improve respective device
Organisms reproduce	Develop self-replicating devices
Organisms coordinate activity through communication	Communication of devices with each other to eliminate abundances
Organisms maintain stability with exquisite control	Feedback mechanisms inside the devices and within the system
Organisms go through natural cycles	Emerging technologies go through circles from primitive to complex to simple
Organisms need emotional satisfaction and intellectual stimulation	Technology should be helpful, and not a burden (cf. openability issues)
Organisms die	Develop materials with expiration date

extract from the natural world, but also on what we can learn from it [20], for example related to developing better brakes. Not only in 1771 this was an issue (see Fig. 3.4), optimizing brakes is still important today.

3.3 Method: The Biomimicry Innovation Method

The biomimicry innovation method (BIM, [21]) is a successful method in biomimetics. This method is applied here to identify biological systems, processes and materials that can inspire tribology. Biomimicry is an innovation method that

Fig. 3.4 1771 crash of Nicolas Joseph Cugnot's steam-powered car into a stonewall. Cugnot was the inventor of the very first self-propelled road vehicle, and in fact he was also the first person to get into a motor vehicle accident

seeks sustainable solutions by emulating nature's time-tested patterns and strategies. The goal is to create products, processes and policies – new ways of living – that are well adapted to life on earth over the long haul.

The steps in BIM are as follows: Identify function, biologize the question, find nature's best practices and generate product ideas.

Identify Function: The biologists distil challenges posed by engineers/natural scientists/architects and/or designers to their functional essence.

Biologize the Question: In the next step, these functions are translated into biological questions such as 'How does nature manage lubrication?' or 'How does nature bond parts together?' The basic question is 'What would nature do here?'

Find Nature's Best practices: Scientific databases as well as living nature itself are used to obtain a compendium of how plants, animals and ecosystems solve the specific challenge.

Generate Process/Product Ideas: From these best practices, the biologists generate ideas for cost-effective, innovative, life-friendly and sustainable products and processes.

The BIM proves highly useful in habitats with high species variety and therefore high innovation potential (e.g. in the tropical rainforests or in corral reefs), providing a multitude of natural models to learn from and emulate. According to the experience of the US based Biomimicry Guild, about 90% of the generated process/product ideas are usually new to their clients (who include companies such as Boeing, Colgate–Palmolive, General Electric, Levi's, NASA, Nike and Procter and Gamble).

There is an abundance of biological literature available. However, only a few of these works concentrate on the functions of biological materials, processes, organisms and systems [19, 22–27]. The Biomimicry Guild is currently undergoing a major endeavour and collects on its web-page http://www.asknature.org 'strategies of nature' together with scientific references and envisaged and already existing bioinspired applications in industry. The 1,245 strategies (status 5 May 2009) are grouped in 8 major sections and comprise answers to the questions

How does nature break down?
How does nature get, store or distribute resources?
How does nature maintain community?
How does nature maintain physical integrity?
How does nature make?
How does nature modify?
How does nature move or stay put?
How does nature process information?

Strategies in 'How does nature maintain community?' of relevance regarding tribology are concerned with maintenance of physical integrity, management of structural forces and prevention of structural failure (Table 3.2). Strategies in 'How does nature move or stay put?' with most relevance regarding tribology are concerned with attachment (Table 3.2). The results section below presents the outcome of a thorough screening of these strategies and subsequent clustering and further analysis of especially promising ones regarding tribology.

3.4 Results: Biomimetics in Tribology – Best Practices and Possible Applications

Application of the BIM concerning wear, shear, tension, buckling, fatigue, fracture (rupture), deformation and permanent or temporal adhesion yields a variety of best practices that comprise biological materials and processes in organisms as diverse as kelp, banana leafs, rattan, diatoms and giraffes (Tables 3.3–3.9).

Table 3.2 Structure of the strategies on AskNature.org relevant for tribology used in this work

Major category	Category	Sub category
Maintain physical integrity (799)	Manage structural forces (289)	Mechanical wear (30)
		Shear (16)
		Tension (28)
	Prevent structural failures (52)	Buckling (14)
		Deformation (4)
		Fatigue (4)
		Fracture (rupture) (30)
Move or stay put (43)	Attach (102)	Permanently (41)
		Temporarily (61)

The numbers indicate the total amount of strategies in the respective categories (status 5 May 2010).

Multifunctionality is a key property in biological entities. Therefore, many organisms and strategies are relevant for more than one tribological issue and therefore also appear in more than just one of the tables given below.

The inspiring organisms, ecosystems and natural structures and functions lay a sound foundation to proceed to the next step: detailed experimental investigation of the phenomena of interest. Further analysis concerning the rich flora in Southeast Asia by one of the authors (ICG) might provide further useful input concerning novel approaches regarding tribology. Valuable literature in this regard is available in abundance [e.g. [28–30]] and personal presence in Malaysia with direct contact to devoted naturalists such as H.S. Barlow with his 96 acres Genting Tea Estate where he plants rare species and provides perfect environment for his objects of study prove highly beneficial for biomimetics work.

Increasing awareness about the innovation potential of the rainforest might also hopefully cause a paradigm shift in the way locals view the pristine forests. With the fast pace people are currently cutting down pristine tropical forests (e.g. in Asia or Brazil) and the subsequent extinction of a multitude of species, many of which are even not yet known to the public, many inspiring plants and animals are lost forever, before we even have started to value them. Gebeshuber and co-workers have recently proposed a niche tourism concept for Malaysia and Thailand, where corporate tourists and local bioscouts practice biomimetics in rainforests, coastal and marine environments and thereby provide sustainable usage of pristine tropical environment, increased income and employment in the host countries while encouraging conservation and sustainable tourism development [31, 32].

3.4.1 Application of the Biomimicry Innovation Method Concerning Mechanical Wear

Wear concerns the erosion of material from a solid surface by the action of another surface. It is related to surface interactions and more specifically the removal of material from a surface as a result of mechanical action. The need for mechanical action, in the form of contact due to relative motion, is an important distinction between mechanical wear and other processes with similar outcomes (e.g. chemical corrosion) [33]. Table 3.3 summarizes the application of the BIM regarding mechanical wear.

The lubrication strategies applied in chameleon tongues could for example be investigated regarding lubrication in bionanotechnological devices and fast actuators.

> 'The chameleon's tongue moves at ballistic speeds – the acceleration reaches 50 g – five times more than an F16 fighter jet. The burst of speed is produced by spiral muscles in the tongue, which contract width-wise to make them stretch forward. A lubricant allows the muscles to slide at time-slicing speeds.' [36, p. 70].

At the core of a chameleon's tongue is a cylindrical tongue skeleton surrounded by the accelerator muscle. High-speed recordings of *Chamaeleo melleri* and

Table 3.3 Application of the biomimicry innovation method regarding mechanical wear

Biologized question: How does nature ...	Nature's best practice	Generated process/product ideas
... build flexible anchors?	Anchor has flexibility: bull kelp [34]	Bioinspired wave and tidal power systems [35]
... lubricate fast moving parts?	Chameleon tongues move with an acceleration of 50 g and are lubricated [36, p. 70]	Lubrication in bionanotechnological devices, fast actuators
... protect seeds from wear?	Seed coat: lotus (*Nelumbo nucifera*) [37]	Packaging
... protects trees from damage?	Resin protects damage: conifer trees [38]	Packaging
... lubricate joints?	Coefficient of friction in hip joints: 0.001 [39–41]	Technical joints, hip implants
... prevent wear in abrasive conditions?	Skin exhibits low friction: sandfish skink [42]; optimized tribosystem: snake skin [43–45]	Abrasive cutting cools, adaptation in plateau honed surfaces [46]
... maintain sharpness of teeth?	Teeth are self-sharpening: American beaver [47], sea urchins [48]	Self-sharpening tools, abrasive cutting cools, self-sharpening hand and power saws [49]
... maintain low friction in nanoscale parts in relative motion?	Moving parts are lubricated: diatoms [50]	3D-MEMS [51]
... protect soft matter against wear?	The skin of cartilaginous fish (*Elasmobranchii*) is protected by a covering of abrasive placoid scales, called denticles [52, p. 91]; skin and mucus prevent abrasion: blennies [53].	Self-sharpening tools, abrasive cutting cools, industrial-grade sanders
... protect bodies from dirt particles?	The body and eyes of stonefly larvae (*Capniidae*) are protected from sediment particles by a coating of dense hairs and bristles [54, p. 115].	Surface layer of devices that come in contact with abrasive particles
... control wear of teeth?	Long-lived grazers with a side-by-side layered arrangement of enamel, dentine and cement [25, p. 333]	Agricultural tools
... protect skin when burrowing?	Webbed feet of the platypus (*Ornithorhynchus anatinus*) are used for burrowing by folding back the webbing to expose the claws for work [55]	Protect equipment from damage, or from damaging something it comes into contact with when not in use. Gloves
... protect folded structures from wear?	Insect wings [56]	Packaging, manufacturing, transport
... protect soft structures from thorns?	Leathery tongue (*Giraffa camelopardalis*) [57, p. 61]	Soft but durable packaging replacing hard plastics

Possible application scenarios are presented in the third column of this table.

C. pardalis reveal that peak powers of 3,000 W/kg are necessary to generate the observed accelerations. The key structure in the projection mechanism is probably a cylindrical connective-tissue layer, which surrounds the entoglossal process and acts as lubricating tissue. Thus, the chameleon utilizes a unique catapult mechanism that is very different from standard engineering designs [58]. Industrial sectors interested in this strategy could be manufacturing, food and medicine; possible application ideas comprise bio-friendly lubrication for use in industry and actuators that lengthen quickly.

3.4.2 Application of the Biomimicry Innovation Method Concerning Shear

Shear concerns a deformation of an object in which parallel planes remain parallel but are shifted in a direction parallel to themselves. In many man-made materials, such as metals or plastics, or in granular materials, such as sand or soils, the shearing motion rapidly localizes into a narrow band known as a shear band. In that case, all the sliding occurs within the band, while the blocks of material on either side of the band simply slide past one another without internal deformation. A special case of shear localization occurs in brittle materials when they fracture along a narrow band. Then, all subsequent shearing occurs within the fracture. Table 3.4 summarizes the application of the BIM regarding shear.

3.4.3 Application of the Biomimicry Innovation Method Concerning Tension

Tension is the magnitude of the pulling force exerted by a string, cable, chain or similar object on another object. It is the opposite of compression. Tension is a force and is always measured parallel to the string on which it is applied. Table 3.5 summarizes the application of the BIM regarding tension.

3.4.4 Application of the Biomimicry Innovation Method Concerning Buckling, Fatigue, Fracture (Rupture) and Deformation

Buckling, fatigue, fracture (rupture) and deformation are well-known phenomena; their specific meaning in tribology is summarized below. *Buckling* is a failure mode characterized by a sudden failure of a structural member subjected to high compressive stresses, where the actual compressive stress at the point of failure

Table 3.4 Application of the biomimicry innovation method regarding shear

Biologized question: How does nature...	Nature's best practice	Generated process/ product ideas
...reinforce materials?	Spiral fibres strengthen tree trunks [59, pp.28–29]: pine; Circular, tapering beams stabilize: plants [60]; Nature achieves high flexural and torsional stiffness in support structures, with minimum material use, by using hollow cylinders as struts and beams [25, p. 440].	Tough materials
...prevent structures from breaking?	Stretchable architecture resists breakage: bull kelp [61]; Joint shaped as suction cup prevents peeling: bull kelp [25, p. 425], Variable postures aid intertidal zone survival: sea palm [25, p. 435]	Tough materials
...build lightweight?	Lightweighting: Scots pine [62]; Bones are lightweight yet strong: birds [63]	Lightweight structures and materials
...resist shear?	Insect elytra resist shear and cracking: beetles [64]; Tissues resist bending under stress: giant green anemone [65]; Pulled support stalks have low flow stress: algae [25, p. 437; 66]; Leaves resist bending: trees, p. 580]; Many organisms, including limpets, resist shearing loads temporarily in part thanks to Stefan adhesion, which occurs when a thin layer of viscous liquid separates two surfaces [25, p. 427].	Shear resistant materials

is less than the ultimate compressive stresses that the material is capable of withstanding. This mode of failure is also described as failure due to elastic instability. Mathematical analysis of buckling makes use of an axial load eccentricity that introduces a moment, which does not form part of the primary forces to which the member is subjected. *Fatigue* is the progressive and localized structural damage that occurs when a material is subjected to cyclic loading. The maximum stress values are less than the ultimate tensile stress limit, and may be below the yield stress limit of the material. *Fracture* mechanics is an important tool in improving the mechanical performance of materials and components. It applies the physics of stress and strain, in particular the theories of elasticity and plasticity, to the microscopic crystallographic defects found in real materials to predict the macroscopic mechanical failure of bodies. *Rupture* or ductile rupture describes the ultimate failure of tough ductile materials loaded in tension. Rupture describes a failure mode in which, rather than cracking, the material 'pulls apart', generally leaving a rough surface. *Deformation* denotes a change in the shape or size of an object due to an applied force. Tables 3.6 and 3.7 summarize the application of the BIM regarding buckling, fatigue and fracture (rupture); and deformation. The

3 Biomimetics in Tribology

Table 3.5 Application of the biomimicry innovation method regarding tension

Nature's best practice	Generated process/ product ideas
Stretchable architecture resists breakage: bull kelp [61]; Stretching mechanism prevents fracture: blue mussel [67]; Two-phase composite tissues handle tension: pipevine [68]; Membranes get fatter when stretched: cells [69]; Arterial walls resist stretch disproportionately: cephalopods [25, pp. 7–8];	Stretchable materials
After too much tension is applied: Bones self-heal: vertebrates [70]; Diatom adhesives self-heal [71]	Self-healing materials; Self-healing coatings [72]
Walls prevent collapse under tension: plants [73]; Fluid pressure provides support: blue crab [74]; Pressure provides structural support: blackback land crab [74]	Reinforcement of foldable structures
Pulled support stalks have low flow stress: algae [25, p. 437; 66]	Construction
Intricate silica architecture ensures mechanical stability under high tension: diatoms [75–77]	MEMS
Crystals and fibres provide strength, flexibility: bones [78]; Byssus threads resist hydrodynamic forces [79]; Silk used for various functions: spiders [80]; Teeth resist compression and tension: animals that chew [25, pp. 332–333]; Elastic ligament provides support, shock absorption: large grazing mammals [25, p. 304]	Tough materials
Circular, tapering beams stabilize: plants [60]; Buttressing resists uprooting: English oak [25, pp. 431–432]; Resisting shearing forces: limpets [25, p. 427]; Variable postures aid intertidal zone survival: sea palm [25, p. 435]; Leaves resist gravitational loading: broad-leaved trees [25, p. 375]; Tentacles maintain tension as flow increases: marine polychaete worm [81]	Stabilize materials
Curved spine deals with tension: sloth [52, p. 37]; Low-energy perching: mousebird [82, pp. 240–241]	Tension resistant materials

biologized question 'How does nature manage changes in humidity?' (Table 3.7, top) is a question resulting from reverse engineering, because we already know that shape change in nature is often initiated by changes in humidity.

3.4.5 Application of the Biomimicry Innovation Method Concerning Attachment

To stay put is important for many organisms; a plenitude of different methods for mechanical attachment or chemical bonding have evolved. In this book chapter, mechanisms to stay put are divided into mechanisms for permanent and temporary attachment.

Table 3.6 Application of the BIM regarding buckling, fatigue and fracture (rupture)

Function	Biologized question: How does nature ... Nature's best practice	Generated process/ product ideas
Buckling	Stems resist buckling: bamboo and other plants [83, 25, p. 378]; Quills resist buckling: porcupine [84]; Siliceous skeleton provides support: Venus flower basket [85]; Shape of feather shafts protect from wind: birds [25, p. 385]; Crystals and fibres provide strength, flexibility: bones [59, p. 32–33; 78]; Organic cases provide protection: bagworm moths [86]; Bones absorb compression shock: birds [52, p. 39]; Leaves resist bending: trees [25, p. 580]; Skeleton provides support: sponges [25, p. 439]; Flexural, torsional stiffness with minimal material use: organisms [25, p. 440]; Spines work as shock absorbers: West European hedgehog [87]; Stems vary stiffness: scouring horsetail [88]	Bioinspired buckling resistant scaffolds
Fatigue	Plants survive repeated drying and rehydration: lesser clubmoss [89]; Wood resists fracture: trees [25, p. 343]; Pulled support stalks have low flow stress: algae [25, p. 437; 66]; Thin 'shells' resist impact loading: sea urchins [25, p. 388; 90–92]; Wings fold multiple times without wear: beetles [56]	Bioinspired fatigue resistant materials
Fracture (rupture)	Bones self-heal: vertebrates [70]; Iron sulphide minerals reinforce scales: golden scale snail [93]; Insect elytra resist shear and cracking: beetles [64]; Tendons and bones form seamless attachment: Chordates [94]; Leaves resist tearing: brown algae [59, pp. 35–36]; Microscopic holes deter fractures: starfish [25, p. 338–339]; Spicules help resist fractures: sponges [25, p. 337]; Extensibility helps stop spread of cracks: macroalgae [25, p. 338; 34, 95]; Shell resists cracking: scallop [25, pp. 339–340]; Leaves resist crosswise tearing: grasses [25, p. 340]; Antlers resist fracture: mammals [25, p. 349]; Resin protects damage: conifer trees [38]; Crystals and fibres provide strength, flexibility: bones [78]; Arterial walls resist stretch disproportionately: cephalopods [25, pp. 7–8]; Hooves resist cracking: horse [96, 97]; Continuous fibres prevent structural weakness: trees [98]; Ctenoid scales form protective layer: bony fish [52, p. 86]; Leaves resist bending: trees [25, p. 580]; Flexural, torsional stiffness with minimal material use: organisms [25, p. 440].	Bioinspired fracture resistant materials

Permanent adhesion can occur via mechanical attachment. One intriguing example for this on the small scale is diatom chains with hinges and interlocking devices that are just some hundreds of nanometers large and that connect the single celled organisms to chains. Some of these connections (still functional) can be found in fossils of diatoms that lived tens of millions of years ago [100]. Most man-made

3 Biomimetics in Tribology

Table 3.7 Application of the BIM regarding deformation

Biologized question: How does nature ...	Nature's best practice	Generated process/product ideas
... manage changes in humidity?	Plants survive repeated drying and rehydration: lesser clubmoss [10, p. 476]	Humidity resistant materials
... build stable scaffolds?	Crystals and fibres provide strength, flexibility: bones [59, pp. 32–33; 78]; Venus flower basket [85]	Scaffold in tissue engineering
... protect soft parts against deformation?	Skin properties derive from arrangement of components: mammals [99]	Mechanical protection (e.g. food packaging)
... provide mechanical stability?	Thin 'shells' resist impact loading: sea urchins [25, p. 388; 90–92]	Hard coated materials

adhesives fail in wet conditions, owing to chemical modification of the adhesive or its substrate. Therefore, bioinspiration from natural underwater adhesives is very much in need. The adhesive that *Eunotia sudetica*, a benthic freshwater diatom species, produces to attach itself to a substrate has for example modular, self-healing properties [50]. Another class of adhesives comprises cement-like materials and adhesives that dry in air. Dry adhesives as they occur in the gecko have been thoroughly investigated, and currently first man-made bioinspired gecko adhesives are produced [101]. Tables 3.8 and 3.9 summarize the application of the BIM regarding permanent and temporary attachment, respectively. In Table 3.9, the mechanical attachment devices for the temporal attachment are structured according to their size (millimetres and above, micrometres and nanometres) – this should help prevent problems with any scaling effect when doing the technology transfer from biology to technology.

Climbing palms, such as the highly specialized rattan palms in the Southeast Asian rainforests, evolved leaves armed with hooks and grapnels for climbing (Fig. 3.5). Some species of rattan palms develop a climbing organ known as the flagellum, which also bears hooks. The leaves are constructed to optimize bending and torsion in relation to the deployment of re-curved hooks. It is a joint phenomenon that hooks in organisms increase in strength toward their base, and that the hooks always fail in strength tests before the part of the organism they are attached to. The sizes and strengths of the hooks differ between species and are related to body size and ecological preference. Larger species produce larger hooks, but smaller climbing palms of the understory deploy fine sharp hooks that are effective on small diameter supports as well as on large branches and trunks. Climbing organs in palms differ significantly from many vines and lianas having more perennial modes of attachment [137].

> 'The front tip, from which all growth comes, explores with extremely long, thin tendrils equipped along their length with needle-sharp curved hooks. If these snag your arm – and the tendrils are so thin that they can easily be overlooked – they can rip both your shirt and

Table 3.8 Application of the BIM regarding permanent attachment

Function	Nature's best practice	Generated process/ product ideas
Permanent adhesion via mechanical attachment	Diatom chains [13, 50, 71, 76]	Hinges and interlocking devices in micromachinery produced via rapid prototyping
Permanent adhesion via wet adhesives	Sticky proteins serve as glue: mammals [102]; Tendons and bones form seamless attachment [63, 78]; Anchor has flexibility: bull kelp [34]; Leaves glued together: grass trees [102]; Mucus glues sand and rock: marine worms [52, pp. 32–33]; Sticky proteins serve as glue: blue mussel [67]; Sticky berries adhere: European mistletoe [103]; Tendrils stick to various surfaces: Virginia creeper [104]; abalone shells [105]	Novel adhesives that can be produced in ambient conditions [106]
Permanent adhesion underwater via wet adhesives	Benthic diatoms [50, 71, 107]	Chemically stable underwater adhesives
Permanent adhesion via cement-like material	Eggs attached securely to hairs with a cement like substance: body lice [108]; Durable casing built with sand: protozoan's [109]; Termite faecal cement [110]	Cement produced at ambient conditions
Permanent adhesion via fluid substances that harden in air or water	Adhesive glues prey: velvet worms [36, p. 78]; Saliva used as glue: swifts [82, p. 239]; Threads adhere underwater: sea cucumber [111]	Novel two component adhesives

your flesh. With these, it hitches itself on to an established tree and actively grows upwards. Sometimes the support is not strong enough to bear the extra load and it collapses, but the rattan is not deterred. It continues to grow as it sprawls across the forest floor and does so with such vigour that some species develop longer stems than any other plant and may reach a length of over five hundred feet.' [57, pp. 162–163]

Bioinspired products and application ideas comprise fasteners, clips, snaps, slide fastener tapes and a novel Velcro analogue (possibly noiseless!) with no need for a counterpart.

3.5 Summary and Outlook

This chapter presented a multitude of best practices from nature concerning meliorated technological approaches of various tribological issues. As next step, detailed investigations on the relevant properties of the best practices shall be performed,

3 Biomimetics in Tribology   41

Table 3.9 Application of the BIM regarding temporary attachment

Temporary adhesion via mechanical attachment devices	Macro to milliscale: spinal column has strength and flexibility: armored shrew (macro) [82, p. 304]; Tendrils enable upward climb: rattan palm [112]; Adhering to multiple substrates: blackberry [57, p. 11]	Velcro analogues with no need for counterparts; novel structures and materials for hanging constructions; novel actuators; attachment of fragile structures
	Microscale: special tongue captures soft prey: long-beaked echidna (*Tachyglossus* = swift tongue; *aculeatus* = furnished with spines) [113]; Insects with two pairs of wings have them work in unison by attaching the wings in various ways, with hooks, folds or catches [114]; Design features aid efficient attachment: lice [115]; Feet grip waxy leaves: leaf beetle [116]; Running on waxy leaves: arboreal ants [25, p. 430; 117]	
	Nanoscale: biological attachment devices from the micro to the nano range [118]	
Temporary adhesion via dry adhesives	Gecko [119, 120]	Dry adhesives [121]
Temporary adhesion via wet adhesives	Mucus takes on adhesive qualities: dusky arion slug [122]; Capillary action aids adhesion: European blowfly [123, 124]; Feet adhere temporarily: aphids [125]	Novel wet adhesives
Temporary adhesion underwater via wet adhesives	Eggs adhere in and out of water: midwife toad [126]; Parasite attaches underwater: copepod [127]; Glue sticks underwater: giant water bug [126, p. 52]; motile diatoms [128]; Adhesive works under water: an aquatic bacterium (nature's strongest glue) [129]	Adhesives that can cure underwater
Temporary adhesion in fluid conditions via switchable adhesives	White blood cells adhere closely: mammals [130]; White blood cells roll and stick: mammals [131]; Sticky berries adhere: European and Australian mistletoe [57, pp. 229–231; 103];	Switchable adhesives (release after signal, adapt binding strength to signal)
	Feet of insects adjust to rough or smooth surfaces by engaging either claws or adhesive foot-pads [115]; Hooks and silk pads aid underwater attachment: blackfly [54, pp. 116–117]; Keyhole limpets attach using either suction or glue-like adhesion [132]; Barnacle cyprids employ wet and dry adhesion [133]; Disk-like structures adhere to smooth surfaces: Spix's disk-winged bat (Stefan and capillary adhesion) [25, p. 427]; Feet grip waxy leaves: leaf beetle [116, 134]; Multiple component glue aids underwater adhesion: barnacle [135]	
Mixed		Bioinspired reversible wet/dry adhesives [136]

Fig. 3.5 Details of the climbing palm rattan. The hooks protect the plant against predators and assist in climbing and growing through the understory in the tropical rain forests. Image reproduced with permission, © F. Saad, IPGM, Malaysia

and the underlying principles shall be extracted. Such principles shall then be incorporated into devices, systems and processes and thereby yield biomimetic technology with increased tribological performance.

To accelerate scientific and technological breakthroughs, we should aim at having a context of knowledge: the gap between scientific insights and technological realization should be bridged [138]. Especially in a field which is as application-oriented as biomimetics related to tribology, care has to be taken that the scientific findings actually can lead to real-world applications. As Gebeshuber and co-workers outlined in 2009 [1] in their 'three gaps theory', there are gaps between inventors, innovators and investors (see Fig. 3.6). 'Inventor gap' denotes the gap between knowing and not knowing that has to be overcome to have ideas. The 'innovator gap' denotes the gap between knowledge and application of the knowledge. The 'investor gap' denotes the gap between the application and the creation of the product. To prevent being trapped in the inventor, innovator or investor gap, a cross dialogue is necessary, a pipeline from 'know-why' to 'know-how' to 'know-what', from the inventor who suggests a scientific or technological breakthrough to the innovator who builds the prototype to the investor who mass produces the product and brings the product to the consumer. Currently, and this is a major problem, at universities worldwide huge amounts of knowledge are piled up with little or no further usage. We know a lot, we can do relatively little. We need a joint language and a joint vision. This is specifically of relevance in tribology, since tribological research is ultimately linked to real-world applications. Applying biomimetics to tribology could provide such a pipeline.

3 Biomimetics in Tribology

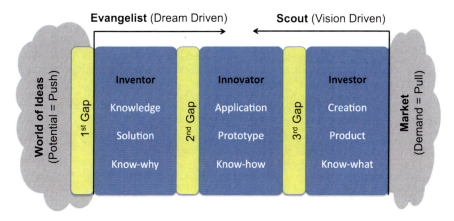

Fig. 3.6 The three gaps theory regarding inventors, innovators and investors. ©2009 PEP Publishing, London. Reproduced from [1] with permission

On the basis of the long-standing experience of research at the interface between tribology and biology [e.g. 2, 8, 12, 13, 14, 100], Gebeshuber and co-workers recently introduced a concept for a dynamic new way of scientific publishing and accessing human knowledge [138]. The authors propose a solution to the dilemma that a plenitude of biology papers that deal with friction, adhesion, wear and lubrication were written solely for a biology readership and have high potential to serve as inspiration for tribology if they were available in a language or in an environment accessible for tribologists (cf. Figs. 3.2 and 3.3).

Acknowledgements The Austrian Society for the Advancement of Plant Sciences has financed part of this work via the Biomimetics Pilot Project 'BioScreen'.

Living in the tropics and continuous exposure to high species diversity in the tropical rainforests is a highly inspirational way to continuously do biomimetics. Researchers have the current problems they are dealing with always at the back of their head, and an inspiring environment aids in developing completely new ideas, approaches and concepts. The Vienna University of Technology, especially Profs. F. Aumayr, H. Störi and G. Badurek, are acknowledged for enabling one of the authors (ICG) three years of research in the inspiring environment in Malaysia.

References

1. I.C. Gebeshuber, P. Gruber, M. Drack, A gaze into the crystal ball – biomimetics in the year 2059. Procd. Inst. Mech. Eng. Part C J Mech Engg Sci **223**(C12), 2899–2918 (2009)
2. I.C. Gebeshuber, B.Y. Majlis, H. Stachelberger, Tribology in Biology: Biomimetic studies across dimensions and across fields. Int. J. Mech. Mat. Engg. **4**(3), 321–327 (2009)
3. B. Bhushan, B.K. Gupta, *Handbook of Tribology: Materials, Coatings, and Surface Treatments* (McGraw-Hill Publishing Company, New York, 1991)
4. G. Stachowiak, A.W. Batchelor, *Experimental Methods in Tribology* (Elsevier B.V., Amsterdam, 2004)

5. C.P. Snow, *The Rede Lecture 1959. The Two Cultures & The Scientific Revolution* (Cambridge University Press, UK, 1959)
6. A. Nordmann, in *Bionik – Aktuelle Forschungsergebnisse in Natur-, Ingenieur- und Geisteswissenschaft*, ed. by T. Rossmann, C. Tropea. Was ist TechnoWissenschaft? – Zum Wandel der Wissenschaftskultur am Beispiel von Nanoforschung und Bionik (Springer, Berlin, 2005) pp. 209–218
7. I.C. Gebeshuber, M. Drack, An attempt to reveal synergies between biology and engineering mechanics. Proc. Inst. Mech. Eng. Part C: J. Mech. Engg. Sci. **222**, 1281–1287 (2008)
8. I.C. Gebeshuber, B.Y. Majlis, L. Neutsch, F. Aumayr, F. Gabor, Nanomedicine and biomimetics: life sciences meet engineering & physics, in *Proceedings of the 3rd Vienna International Conference on Micro- and Nanotechnology*, Viennano09, Editors, 2009, 17–23
9. B. Bhushan, Biomimetics: lessons from nature-an overview. Phil. Transac. Roy. Soc. A. **367**, 1445–1486 (2009)
10. Y. Bar-Cohen, *Biomimetics: Biologically Inspired Technologies* (CRC Press, Boca Raton, FL, 2005)
11. I.C. Gebeshuber, M. Drack, M. Scherge, Tribology in biology. Tribology **2**(4), 200–212, (2008)
12. I.C. Gebeshuber, A. Pauschitz, F. Franek, *Biotribological model systems for emerging nanoscale technologies*, in *Proceedings of 2006 IEEE Conference on Emerging Technologies – Nanoelectronics*, Editors, 2006, 396–400
13. I.C. Gebeshuber, H. Stachelberger, B.A. Ganji, D.C. Fu, J. Yunas, B.Y. Majlis, Exploring the innovational potential of biomimetics for novel 3D MEMS. Adv. Mat. Res. **74**, 265–268 (2009)
14. I.C. Gebeshuber, H. Stachelberger, M. Drack, in *Life Cycle Tribology*, Tribology and Interface Engineering Series 48, ed. by B.J. Briscoe. Diatom Tribology (Elsevier, Amsterdam, 2005) pp. 365–370
15. A.T. Johnson, *Biology for engineers*. In preparation
16. P. Fratzl, R. Weinkamer, Nature's hierarchical materials. Prog. Mat. Sci. **52**(8), 1263–1334 (2007)
17. J.F.V. Vincent, Deconstructing the design of a biological material. J. Theor. Biol. **236**, 73–78 (2005)
18. W. Nachtigall, *Vorbild Natur: Bionik-Design für funktionelles Gestalten* (Springer, Berlin, 1997)
19. W. Nachtigall, *Das große Buch der Bionik* (Deutsche Verlagsanstalt, Germany, 2003) p. 214f
20. http://www.asknature.org/article/view/what_is_biomimicry
21. Biomimicry Innovation Method (Biomimicry Guild, Helena, MT, USA, 2008)
22. S. Vogel, *Life's Devices: The Physical World of Animals and Plants* (Princeton University Press, Princeton, 1988)
23. S. Vogel, *Life in Moving Fluids: The Physical Biology of Flow* (Princeton University Press, Princeton, 1996)
24. S. Vogel, *Cats' Paws and Catapults: Mechanical Worlds of Nature and People* (WW Norton & Co, New York, 1998)
25. S. Vogel, *Comparative Biomechanics: Life's Physical World* (Princeton University Press, Princeton, 2003)
26. S. Vogel, *Nature's Flyers: Birds, Insects, and the Biomechanics of Flight* (The Johns Hopkins University Press, USA, 2004)
27. M. Scherge, S.S.N. Gorb, *NanoScience and Technology* (Springer, Germany, 2001)
28. A.R. Wallace, *The Malay Archipelago (Stanfords Travel Classics)* (John Beaufoy Publishing Ltd., Oxford, UK, 2008)
29. H.S. Barlow, *An introduction to the moths of South East Asia* (Malaysian Nature Society, Kuala Lumpur, Malaysia, 1982)
30. H.F. Macmillan (revised by H.S. Barlow, I. Enoch, R.A. Russell) *H.F. Macmillan's tropical planting and gardening* (Malaysian Nature Society, Kuala Lumpur, Malaysia, 1991)

31. I.C. Gebeshuber, B.Y. Majlis, *3D corporate tourism: A concept for innovation in nanomaterials engineering*. Int. J. Mat. Engg. Innov. **2**(1) 38–48. Online ISSN 1757-2762, Print ISSN 1757-2754.
32. R. Esichaikul, M. Macqueen, I.C. Gebeshuber, 3D corporate tourism: Application-oriented problem solving in tropical rainforests, in *Proceedings 8th Asia-Pacific CHRIE Conference*, Pattaya, Thailand, 3–6 June 2010, in press
33. J.A. Williams, Wear and wear particles – Some fundamentals. Tribol. Int. **38**(10), 863–870 (2005)
34. M. Denny, B. Gaylord, *The mechanics of wave-swept algae*. J. Exp. Biol. **205**, 1355–1362 (2002)
35. Australian company BioPower Systems Pty. Ltd. (http://www.biopowersystems.com/, Accessed 3 May 2010
36. J. Downer, *Weird Nature: An Astonishing Exploration of Nature's Strangest Behavior* (Firefly Books, Ontario, Canada, 2002)
37. J. Shen-Miller, Sacred lotus, the long-living fruits of China Antique. Seed Sci. Res. **12**(03) 131–143, (2007)
38. N. Kamata, in *Mechanisms and deployment of resistance in trees to insects*, ed. by M.R. Wagner, K.M. Clancy, F. Lieutier, T.D. Paine. Deployment of Tree Resistance to Pests in Asia (Kluwer Academic Publishers, Dordrecht, NL, 2002) p. 277
39. F.C. Linn, Lubrication of animal joints. I. The arthrotripsometer. J. Bone Joint Surg. Am. **49**, 1079–1098 (1976)
40. C.W. McCutchen, Mechanism of animal joints: sponge-hydrostatic and weeping bearings. Nature **184**, 1284–1285 (1959)
41. A. Unsworth, D. Dowson, V. Wright, The frictional behavior of human synovial joints – Part I: natural joints. Trans. Am. Soc. Mech. Eng. F 97F, 369–376 (1975)
42. W. Baumgartner, F. Saxe, A. Weth, D. Hajas, D. Sigumonrong, J. Emmerlich, M. Singheiser, W. Bohme, J.M. Schneider, The sandfish's skin: morphology, chemistry and reconstruction. J. Bionic Eng. **4**(1), 1–9 (2007).
43. H.A. Abdel-Aal, M. El Mansori, S. Mezghani, Multi-scale investigation of surface topography of Ball Python (Python regius) shed skin in comparison to human skin. Tribol. Lett. **37**(3), 517–527 (2010)
44. H.A. Abdel-Aal, M. El Mansori, S. Mezghani, Multi scale surface characterization of shed ball python skins compared to human skin. Tribol. Lett. **37**(3), 517–527 (2010)
45. H.A. Abdel-Aal, M. El Mansori, in *'Biomimetics – Materials, Structures and Processes. Examples, Ideas and Case Studies'*, ed. by Bruckner D., Gruber P., Hellmich C., Schmiedmayer H.-B., Stachelberger H., Gebeshuber I.C. *Python regius* (Ball Python) shed skin: Biomimetic analogue for function-targeted design of tribo-surfaces. Series: Biological and Medical Physics, Biomedical Engineering, Series Editor Claus Ascheron. (Springer Publishing, NY, 2009) this volume
46. H.A. Abdel-Aal, M. El Mansori, Multi scale characterization of surface topography of ball python (Python Regius) shed skin for possible adaptation in plateau honed surfaces. Wear, submitted
47. R.M. Joeckel, A functional interpretation of the masticatory system and paleoecology of entelodonts. Paleobiol. **16**(4), 459–482 (1990)
48. Y. Maa, B. Aichmayer, O. Paris, P. Fratzl, A. Meibom, R.A. Metzler, Y. Politi, L. Addadi, P.U.P.A. Gilbert, S. Weiner, The grinding tip of the sea urchin tooth exhibits exquisite control over calcite crystal orientation and Mg distribution. Proc. Natl. Acad. Sci **106**(15), 6048–6053 (2009)
49. H.H. Payson, *Keeping the Cutting Edge: Setting and Sharpening Hand and Power Saws* (Wooden Boat Publications, Maine, USA, 1985)
50. I.C. Gebeshuber, J.H. Kindt, J.B. Thompson, Y. Del Amo, H. Stachelberger, M. Brzezinski, G.D. Stucky, D.E. Morse, P.K. Hansma, Atomic force microscopy study of living diatoms in ambient conditions. J. Microsc. **212**(Pt3), 292–299 (2003)

51. I.C. Gebeshuber, H. Stachelberger, B.A. Ganji, D.C. Fu, J. Yunas, B.Y. Majlis, Exploring the innovational potential of biomimetics for novel 3D MEMS. Adv. Mat. Res. **74**, 265–268 (2009)
52. S. Foy, *Grand Design: Form and Colour in Animals* (BLA Publishing Limited for J.M.Dent & Sons Ltd, Aldine House, London, UK, 1982)
53. M.H. Horn, R.N. Gibson, in *Life at the Edge: Readings from Scientific American Magazine*, ed. by J.L. Gould, C.G. Gould. Intertidal Fishes (Freeman W.H. and Co., New York, USA, 1989), 61–62
54. P.S. Giller, B. Malmqvist, *The Biology of Streams and Rivers* (Oxford University Press, USA, 1998)
55. D. Attenborough, *Life on Earth* (Little, Brown and Company, Boston, MA, 1979) p. 204
56. F. Haas, Evidence from folding and functional lines of wings on inter-ordinal relationships in Pterygota. Arthropod Systemat. Phylog. 64(2), 149–158 (2006)
57. D. Attenborough, *The Private Life of Plants: A Natural History of Plant Behavior* (BBC Books, London, UK, 1995)
58. J.H. de Groot, J.L. van Leeuwen, Evidence for an elastic projection mechanism in the chameleon tongue. Proc. Biol. Sci. **271**(1540), 761–770 (2004)
59. H. Tributsch, *How Life Learned to Live* (The MIT Press, Cambridge, MA, USA, 1984)
60. A. Rosales, G. Morales, D. Green, Mechanical properties of Guatemala pine 2 by 4's. Forest Prod. J. **45**(10), 81–84 (1995)
61. M.A.R. Koehl, S.A. Wainwright, Mechanical adaptations of a giant kelp. Limnol. Oceanogr. **22**, 1067–1071 (1977)
62. C. Mattheck, Teacher tree: the evolution of notch shape optimization from complex to simple. Eng. Fract. Mech. **73**, 1732–1742 (2006)
63. J.D. Currey, *Bones: Structure and Mechanics* (Princeton University Press, Princeton, USA, 2006)
64. J. Fan, B. Chen, Z. Gao, C. Xiang, Mechanisms in failure prevention of bio-materials and bio-structures. Mech. Adv. Mat. Struct. **12**(3), 229–237 (2005)
65. M.A.R. Koehl, Mechanical organization of cantilever like sessile organisms: sea anemones. J. Exp. Biol. **69**(1), 127–142 (1977)
66. M.A.R. Koehl, How do benthic organisms withstand moving water. Amer. Zoologist. **24**, 57–70 (1984)
67. G.E. Fantner, E. Oroudjev, G. Schitter, L.S. Golde, P. Thurner, M.M. Finch, P. Turner, T. Gutsmann, D.E. Morse, H. Hansma, Sacrificial bonds and hidden length: unraveling molecular mesostructures in tough materials. Biophys. J. **90**(4), 1411–1418 (2006)
68. L. Kohler, H.C. Spatz, Micromechanics of plant tissues beyond the linear-elastic range. Planta. **215**(1), 33–40 (2002)
69. R. Lakes, A broader view of membranes. Nature **414**, 503–504 (2001)
70. E. Canalis, Notch signaling in osteoblasts. Sci. Signal. **1**(17), pe17 (2008)
71. I.C. Gebeshuber, J.B. Thompson, Y. Del Amo, H. Stachelberger, J.H. Kindt, In vivo nanoscale atomic force microscopy investigation of diatom adhesion properties. Mat. Sci. Technol. **18**, 763–766 (2002)
72. A. Kumar, L.D. Stephenson, J.N. Murray, Self-healing coatings for steel. Progr. Org. Coat. **55**(3), 244–253 (2006)
73. D.F. Cutler, in *Nature and Design*, ed. by M.W. Collins, M.A. Atherton, J.A. Bryant. Design in Plants (WIT Press, Southampton, Boston, USA, 2005) pp. 99–100
74. J.R.A. Taylor, W.M. Kier, A pneumo-hydrostatic skeleton in land crabs. Nature **440**(7087), 1005 (2006)
75. C.E. Hamm, R. Merkel, O. Springer, P. Jurkojc, C. Maier, K. Prechtel, V. Smetacek, Architecture and material properties of diatom shells provide effective mechanical protection. Nature **421**(6925), 841–843 (2003)
76. I.C. Gebeshuber, R.M. Crawford, Micromechanics in biogenic hydrated silica: hinges and interlocking devices in diatoms. Proc. Inst. Mech. Eng. Part J: J. Engg. Tribol. **220**(J8), 787–796 (2006)

77. R.M. Crawford, I.C. Gebeshuber, Harmony of beauty and expediency. Sci. First Hand **5**(10), 30–36 (2006)
78. M.J. Buehler, Molecular nanomechanics of nascent bone: fibrillar toughening by mineralization. Nanotechnology **18**(29), 295102(9p) (2007)
79. M.W. Denny, *Biology and the Mechanics of the Wave-Swept Environment* (Princeton University Press, Princeton, USA, 1988)
80. M. Xu, R.V. Lewis, Structure of a protein superfiber: spider dragline silk. Proc. Natl. Acad. Sci. USA **87**(18), 7120–7124 (1990)
81. A.S. Johnson, Sag-mediated modulated tension in terebellid tentacles exposed to flow. Biol. Bull. **185**, 10–19 (1993)
82. M.E. Fowler, R.E. Miller, *Zoo and Wild Animal Medicine* (W.B. Saunders Co., Philadelphia, USA 2003)
83. K. Schulgasser, A. Witztum, On the strength, stiffness, and stability of tubular plant stems and leaves. J. Theor. Biol. **155**, 497–515 (1992)
84. G.N. Karam, L.J. Gibson, Biomimicking of animal quills and plant stems: natural cylindrical shells with foam cores. Mat. Sci. Engg C **2**(1–2), 113–132 (1994)
85. J.C. Weaver, J. Aizenberg, G.E. Fantner, D. Kisailus, A. Woesz, P. Allen, K. Fields, M.J. Porter, F.W. Zok, P.K. Hansma, P. Fratzl, D.E. Morse, Hierarchical assembly of the siliceous skeletal lattice of the hexactinellid sponge *Euplectella aspergillum*. J. Struct. Biol. **158**(1), 93–106 (2007)
86. E. Tsui, *Evolutionary Architecture: Nature as a Basis for Design* (Wiley, New York, 1999) p. 128
87. J.F.V. Vincent, Survival of the cheapest. Mat. Today **5**(12), 28–41 (2002)
88. T. Speck, T. Masseleter, B. Prum, O. Speck, R. Luchsinger, S. Fink, Plants as concept generators for biomimetic lightweight structures with various stiffness and self-repair mechanisms. J. Bionics Engg. **1**(4), 199–205 (2004)
89. Y. Bar-Cohen, *Biomimetics: Biologically Inspired Technologies* (CRC/Taylor & Francis, Boca Raton, FL, USA, 2006)
90. M. Telford, Domes, arches and urchins: the skeletal architecture of echinoids (Echinodermata). Zoomorphology **105**, 125–134 (1985)
91. O. Ellers, M. Telford, Causes and consequences of fluctuating coelomic pressure in sea urchins. Biol. Bull. **182**(3), 424–434 (1992)
92. O. Ellers, A.S. Johnson, P.E. Mober, Structural strengthening of urchin skeletons by collagenous sutural ligaments. Biol. Bull. **195**(2), 136–144 (1998)
93. M. Schrope, Deep sea special: The undiscovered oceans. New Scientist **188**(2525), 36–43 (2005)
94. R.B. Martin, D.B. Burr, N.A. Sharkey, *Skeletal Tissue Mechanics* (Springer, NY, 1998) pp. 330–332
95. K.J. Mach, B.B. Hale, M.W. Denny, D.V. Nelson, Death by small forces: a fracture and fatigue analysis of wave-swept macroalgae. J. Exp. Biol. **210**(13), 2231–2243 (2007)
96. M.A. Kasapi, J.M. Gosline, Design complexity and fracture control in the equine hoof wall. J. Exp. Biol. **200**, 1639–1659 (1997)
97. M.A. Kasapi, J.M. Gosline, Micromechanics of the equine hoof wall: optimizing crack control and material stiffness through modulation of the properties of keratin. J. Exp. Biol. **202**, 377–391 (1999)
98. U. Luettge, *Physiological Ecology of Tropical Plants* (Springer, Berlin, 2007)
99. L. Ambrosio, P.A. Netti, L. Nicolais, in *Integrated biomaterials science*, Ed. by R. Barbucci. Tissues (Kluwer Academic Publishers, New York, 2002) p. 356
100. I.C. Gebeshuber, Biotribology inspires new technologies. Nano Today **2**(5), 30–37 (2007)
101. P. Forbes, *The Gecko's Foot: Bio-Inspiration: Engineering New Materials from Nature* (W.W. Norton & Company, New York, USA, 2006)
102. J.D. Smart, The basics and underlying mechanisms of mucoadhesion. Adv. Drug Deliv. Rev. **57**(11), 1556–1568 (2005)

103. J.-I. Azuma, N.-H. Kim, L. Heux, R. Vuong, H. Chanzy, The cellulose system in viscin from mistletoe berries. Cellulose **7**(1) 3–19 (2000)
104. A.J. Bowling, K.C. Vaughn, Gelatinous fibers are widespread in coiling tendrils and twining vines. Am. J. Bot. **96**, 719–727 (2009)
105. B.L. Smith, T.E. Schäffer, M. Viani, J.B. Thompson, N.A. Frederick, J. Kindt, A. Belcher, G.D. Stucky, D.E. Morse, P.K. Hansma, Molecular mechanistic origin of the toughness of natural adhesives, fibres and composites. Nature **399**, 761–763 (1999)
106. H. Lee, S.M. Dellatore, W.M. Miller, P.B. Messersmith, *Mussel-inspired surface chemistry for multifunctional coatings*. Science **318**(5849), 426–430 (2007)
107. T.M. Dugdale, R. Dagastine, A. Chiovitti, P. Mulvaneyand, R. Wetherbee, Single adhesive nanofibers from a live diatom have the signature fingerprint of modular proteins. Biophys. J. **89**(6), 4252–4260 (2005)
108. W.S. Pray, Head lice: perfectly adapted human predators. Am. J. Pharm. Edu. **63**, 204–209 (1999)
109. D.J.G. Lahr, S.G.B.C. Lopes, Morphology, biometry, ecology and biogeography of five species of *Difflugia Leclerc, 1815 (Arcellinida: Difflugiidae)*, from Tiete River, Brazil. Acta Protozoologica **45**, 77–90 (2006)
110. A.M. Stuart, Alarm, defense, and construction behavior relationships in termites *(Isoptera)*. Science **156**, 1123–1125 (1967)
111. P. Flammang, J. Ribesse, M. Jangoux, Biomechanics of adhesion in sea cucumber cuvierian tubules (Echinodermata, Holothuroidea). Integ. Comp. Biol. **42**(6), 1107–1115 (2002)
112. D. Attenborough, *The Private Life of Plants: A Natural History of Plant Behavior* (BBC Books, London, 1995) pp. 162–163
113. G.A. Doran, H. Baggett, The specialized lingual papillae of *Tachyglossus aculeatus* I. Gross and light microscopic features. Anatom. Rec. **172**(2), 157–165 (1971)
114. A. Wootton, *Insects of the World* (Blandford Press, Poole, 1984) p. 36
115. S. Gorb, *Attachment Devices of Insect Cuticle* (Kluwer Academic Publishers, Dortrecht, 2001) pp. 56, 60–61
116. S.N. Gorb, R.G. Beutel, E.V. Gorb, Y. Jiao, V. Kastner, S. Niederegger, V.L. Popov, M. Scherge, U. Schwarz, W. Votsch, Structural design and biomechanics of friction-based releasable attachment devices in insects. Integ. Comp. Biol. **42**(6) 1127–1139 (2002)
117. W. Federle, K. Rohrseitz, B. Hölldobler, Attachment forces of ants measured with a centrifuge: better 'wax-runners' have a poorer attachment to a smooth surface. J. Exp. Biol. **203**(3), 505–512 (2000)
118. E. Arzt, S. Gorb, R. Spolenak, From micro to nano contacts in biological attachment devices. Proc. Natl. Acad. Sci. USA **100**(19), 10603–10606 (2003)
119. K. Autumn, Y.A. Liang, S.T. Hsieh, W. Zesch, W.P. Chan, T.W. Kenny, R. Fearing, R.J. Full, Adhesive force of a single gecko foot-hair. Nature **405**(6787), 681–685 (2000)
120. Y. Tian, N. Pesika, H. Zeng, K. Rosenberg, B. Zhao, P. McGuiggan, K. Autumn, J. Israelachvili, Adhesion and friction in gecko toe attachment and detachment. Proc. Natl. Acad. Sci. USA **103**(51), 19320–19325 (2006)
121. A. Parness, D. Soto, N. Esparza, N. Gravish, M. Wilkinson, K. Autumn, M. Cutkosky, A microfabricated wedge-shaped adhesive array displaying gecko-like dynamic adhesion, directionality and long lifetime. J. Roy. Soc. Int/Roy. Soc. **6**, 1223–1232 (2009)
122. J.M. Pawlicki, L.B. Pease, C.M. Pierce, T.P. Startz, Y. Zhang, A.M. Smith, The effect of molluscan glue proteins on gel mechanics. J. Exp. Biol. **207**, 1127–1135 (2004)
123. M. Hopkin, Flies get a grip. Nature **431**(7010), 756–756 (2004)
124. M.G. Langer, J.P. Ruppersberg, S. Gorb, Adhesion forces measured at the level of a terminal plate of the fly's seta. Proc. Roy. Soc. B: Biol. Sci. **271**(1554), 2209–2215 (2004)
125. A.F.G. Dixon, P.C. Croghan, R.P. Gowing, The mechanism by which aphids adhere to smooth surfaces. J. Exp. Biol. **152**(1), 243–253 (1990)
126. M. Crump, *Headless Males Make Great Lovers and Other Unusual Natural Histories* (University of Chicago Press, Chicago, IL, USA, 2005) p. 53

127. A.L. Ingram, A.R. Parker, The functional morphology and attachment mechanism of pandarid adhesion pads (Crustacea: Copepoda: Pandaridae). Zoologischer Anzeiger – J. Comp. Zool. **244**(3–4), 209–221 (2006)
128. M. Drack, L. Ector, L.C. Gebeshuber, R. Gordon, in *The Diatom World*, ed. by J. Seckbach, J.P. Kociolek. A review of diatom gliding motility, with a new action potential theory for its control and potential uses in open channel active nanofluidics (Springer, Dordrecht, The Netherlands, 2009) in preparation
129. P.H. Tsang, G. Li, Y.V. Brun, L.B. Freund, J.X. Tang, Adhesion of single bacterial cells in the micronewton range. Proc. Natl. Acad. Sci. USA **103**, 5764–5768 (2006)
130. D.F. Tees, D.J. Goetz, Leukocyte adhesion: an exquisite balance of hydrodynamic and molecular forces. News Physiol. Sci. **18**, 186–90 (2003)
131. C.E. Orsello, D.A. Lauffenburger, D.A. Hammer, Molecular perspectives in cell adhesion: a physical and engineering perspective. Trends Biotechnol. **19**, 310–316 (2001)
132. A.M. Smith, Alternation between attachment mechanisms by limpets in the field. J. Exp. Marin. Biol. Ecol. **160**(2) 205–220 (1992)
133. Y. Phang, N. Aldred, A.S. Clare, G.J. Vancso, Towards a nanomechanical basis for temporary adhesion in barnacle cyprids (*Semibalanus balanoides*). J. Roy. Soc. Int/Roy. Soc. **5**(21), 397–402 (2008)
134. T. Eisner, *For Love of Insects* (The Belknap Press of Harvard University Press, Cambridge, MA, USA, 2005) pp. 134–135
135. K. Kamino, Novel barnacle underwater adhesive protein is a charged amino acid-rich protein constituted by a Cys-rich repetitive sequence. Biochem. J. **356**(2), 503–507 (2001)
136. H. Lee, B.P. Lee, P.B. Messersmith, A reversible wet/dry adhesive inspired by mussels and geckos. Nature **448**, 338–341 (2007)
137. S. Isnard, N.P. Rowe, The climbing habit in palms: Biomechanics of the cirrus and flagellum. Am. J. Bot. **95**, 1538–1547 (2008)
138. I.C. Gebeshuber, B.Y. Majlis, New ways of scientific publishing and accessing human knowledge inspired by transdisciplinary approaches. Trib. Surf. Mat. Int. **4**(3), 143–151 (2010)

Chapter 4
Reptilian Skin as a Biomimetic Analogue for the Design of Deterministic Tribosurfaces

H.A. Abdel-Aal and M. El Mansori

Abstract A major concern in designing tribosystems is to minimize friction, save energy, and to reduce wear. Satisfying these requirements depends on the integrity of the rubbing surface and its suitability to sliding conditions. As such, designers currently focus on constructing surfaces that are an integral part of the function of the tribosystem. Inspirations for such constructs come from studying natural systems and from implementing natural design rules. One species that may serve as an analogue for design is the Ball Python. This is because such a creature doesn't sustain much damage while depending on legless locomotion when sliding against various surfaces, many of which are deemed tribologically hostile. Resistance to damage in this case originates from surface design features. As such, studying these features and how do they contribute to the control of friction and wear is very attractive for design purposes. In this chapter, we apply a multiscale surface characterization approach to study surface design features of the *Python regius* that are beneficial to design high-quality lubricating surfaces (such as those obtained through plateau honing). To this end, we studied topographical features by SEM and through white light interferrometery. We further probe the roughness of the surface on multiscale and as a function of location within the body. The results are used to draw a comparison to metrological features of commercial cylinder liners obtained by plateau honing.

H.A. Abdel-Aal (✉)
Arts et Métier Paris Tech, Rue Saint Dominique BP 508, 51006 Chalons-en-Champagne Cedex, France
e-mail: hisham.abdel-aal@chalons.ensam.fr

4.1 Introduction

One of the current pressing technical problems facing engineers is to curb global consumption of fossil fuels. At the core of the strategy devised to achieve such a goal is to reduce the energy consumption footprint of internal combustion engines (ICEs). It is to be noted that most of the fuel savings arise from reducing energy losses to friction resistance to motion. ICEs typically operate with a thermal efficiency, ratio of output energy to input energy, which falls between 50% and 60% [1]. It is estimated [2] that roughly 15% of the energy input to a passenger vehicle is consumed by friction. In the United States alone, more than two hundred million motor vehicles are powered by ICEs. Many of these engines produce power in the order of 10^2 kW. So that, friction-induced power losses translate into a staggering amount of wasted crude oil (estimated to be in excess of one million barrels a day in the United States alone).

Frictional forces in ICEs are a consequence of hydrodynamic stresses in oil films and metal to metal-to-metal contacts. Friction-induced power losses represent a significant fraction of the overall power produced by an ICE. Piston–cylinder friction contributes a fare share of such lost power. Consequently, minimization of the friction losses due to the operation of these components is a major concern for engine designers. Typically, ICEs are lubricated to minimize the friction between moving parts. Lubrication, however, does not eliminate friction, and it may cause significant damage to the engine on failure. Most of the frictional losses within the moving parts of an ICE result from friction between piston and cylinder. Because of the pressures acting on the piston and the cylinder, hydrodynamic stresses will develop in the oil films. These may contribute to frictional losses through fluid friction resulting from viscosity effects. Additionally, an engine cylinder will contain very hot gases resulting from the ignition of fuel. The high temperatures encountered may lead to failure of the lubrication layer and hence the occurrence of metal-to-metal contact between cylinder and piston. Such a situation is not desirable since it leads to engine seizure.

Several factors affect the energy consumed in friction between solids. These may be broadly classified as intrinsic factors that pertain to the physical and chemical properties of the rubbing materials, and extrinsic factors that relate to applied loads, ambient operation temperatures, rubbing speeds, and topography of the rubbing surfaces, and so on. Within the extrinsic factor group, the topography of the rubbing surfaces crucially influences the integrity of the rubbing pair. This is because the surface of a rubbing interface manifests the changes that a solid undergoes during the manufacturing stages. It also reflects the response of the contacting solids to other extrinsic factors. More importantly, the topography of the rubbing interface affects the quality of lubrication. Therefore, constructing a surface of a predetermined topography, which yields a predictable response, and in the meantime self-adapts in response to changes in sliding conditions, can reduce frictional losses. Such surfaces termed here as "deterministic surfaces" are starting to emerge in many modern engines. These designs are artificial textures that ornament the inner surface of cylinder bores (cylinder liner). There are several methods used to emboss the artificial textures (e.g., multistep honing, helical honing, controlled thin-layer deposition, and laser texturing) [3–12]. To date, however, there is no agreement on the optimal topology that such surfaces should acquire. Furthermore, a systematic methodology that, if applied, may generate deterministic surface designs, which meet functional requirements of a given engine, is virtually nonexistent. Reasons for such a condition are rooted in the analytical philosophy that inspires design generation in man-engineered systems (MES) in comparison to the holistic approach that generates design in natural systems.

Generation of design in natural systems (geometry, pattern, form, and texture) is a holistic phenomenon that synchronizes all design constituents toward an overall optimized performance envelope. Such an approach yields deterministic design outputs that while conceptually simple are of optimized energy expenditure footprint. Natural engineering, thus, seeks transdisciplinary technically viable alternatives that, given functional constraints, require minimum effort to construct and economizes effort while functioning. In nature, there are many examples of designs and technical solutions that economize the effort needed for operation and minimize damage profiles [10]. An analogous design paradigm, within the *MES* domain, has not matured as of yet. In addition, the process of texturing is a multistep operation of a high degree of functional complexity. This presents considerable challenges to surface designers who conceptually conform to conventional paradigms that do not acknowledge functional complexity to start with.

Confining the discussion to texturing by honing, we may identify two sets of interrelated parameter groups within textures of cylinder bores: product functionality factors and a so-called cylinder liner features. Oil consumption, for example, falls under the first parameter group. Each constituent of a parameter group, in turn, relates to a feature factor of the honing process itself. In our example, oil consumption relates to the honing tool angle of application, any chatter marks, residual grooves, and cracks present within the virgin surface. The running-in performance of the cylinder–piston assembly originates from the mechanics of the intermediate step of the texturing process (the so-called plateaux forming stage) and whether foreign bodies are present in the material of the surface to start with. In all, there are 400 parameters involved in artificial texturing of cylinder bores by honing. Consideration of all parameters is considerably difficult if approached by conventional means. To this end, it is necessary to devise a new alternative that permits by passing detailed consideration of the complexity of the process. Additionally, this alternative approach should allow harnessing the complexities of texturing to generate a deterministic design output in the natural sense. That is a surface that self-adapts to changing external contact conditions, and may be able to optimize the tribological response based on intrinsic features rather than external modifiers. To devise the alternative approach one has to define precisely the function of the desired surface. This, in principal, determines the objective of the design process and assists envisioning the final output.

The ultimate function of cylinder bore texturing is to reduce friction between the piston and cylinder walls while reducing the friction stresses on the oil film used in lubrication. This goal is mainly achieved through optimized topographical features of the rubbing surfaces. The mechanistic principle that aids in friction reduction is minimization of contact between cylinder and piston. Once minimization takes place, the overall frictional forces reduce, and damage reduces as well. However, due to sliding, wear will take place. This will alter the contact conditions because of the induced changes in topography. The required surface has to maintain optimal sliding performance despite topography modifications. That is, the texturing has to self-adapt to induced changes in sliding conditions. Through this adaptation, the surface will maintain optimal performance within an envelope of conditions rather than achieving peak performance at a point. Such a requirement is a characteristic of natural systems. As such, inspiration for the envisioned texture has to originate from a natural system analogue. Squamate reptiles are major inspirations in that context. They present diverse examples where surface structuring, and modifications through submicron and nanoscale features, achieves frictional regulation manifested in reduction of adhesion [11], abrasion resistance [12], and frictional anisotropy [13].

Squamata comprises two large clades, Iguania (about 1,230 known species) and Scleroglossa (about 6,000 known species), 3,100 of which are traditionally referred to as "lizards," and the remaining 2,900 species as "snakes" [14]. Squamates have a wide distribution all over the planet. They are found almost everywhere on earth except where factors that limit their survivability are present (e.g., higher altitudes where very cold temperatures are predominant year round). Their ecological diversity, and thereby their diverse habitat, presents a broad range of tribological environments, many of which are hostile in terms of sliding and contact conditions. Such a situation requires specific tribological response that manifests itself in functional practices and surface design features. As such, squamates offer a great resource that can be mined for viable surface design inspirations, which address a wide spectrum of technical problems, faced in MES. Many studies describe appearance and structure of skin in Squamata [15–18]. Studies also describe the geometrical features and the evolution of functional adaptability of many species [19]. Tribological performance of the snake clade was also a subject of many studies in biology, herpetology, and engineering [20]. Researchers have studied the mechanical behavior of snakeskin [21, 22]. Design of bioinspired robots inspired several investigations of snakes to understand the mechanisms responsible for regulating legless locomotion [23]. The design of light-weight high-resolution infrared sensors prompted the study of light detection in vipers [24]. Hazel et al. [13] probed some of the nanoscale design features of three snake species. The authors documented the asymmetric features of the skin ornamentation to which both authors attributed frictional anisotropy. Shafiei and Alpas [25, 26] reported that snakeskin replicas provide anisotropic tribological properties that minimize frictional interaction. Such an effect, these authors argue, stems from the asymmetric shape of the protrusions at the ridges of the skin's scales. Functional adaptation and morphing of snake ornamentation was also subject to several studies. The results attribute adaptation to the presence of submicron- and nanosized fibril structures acting to modify friction and adhesion during locomotion.

4 Reptilian Skin as a Biomimetic Analogue

Snakes are limbless animals. They have multimodes of motion (slithering, crawling, serpentine movement, etc.) that take place during propulsion. Such motion modes are initiated through muscular activity, that is, through a sequence of contraction and relaxation of appropriate muscle groups. Transfer of motion between the body of the snake and the substrate depends on generation of sufficient tractions. This process, generation of traction and accommodation of motion, takes place through the skin. Thus, the skin of the snake assumes the role of motion transfer and accommodation of energy consumed during the initiation of motion. The number, type, and sequence of muscular groups responsible for the initiation of motion, and thus employed in propulsion, will vary according to the particular mode of motion initiated. It will also depend on the habitat and the surrounding environment. This will also affect the effort invested in initiation of motion and thereby affects the function of the different parts of the skin and the amount of accommodated energy. Therefore, in general, different parts of the skin will have different functional requirements. Moreover, the life habits of the particular species, for example, defense, hunting, and swallowing) will require different deterministic functions of the different parts of the skin.

In a recent study, Abdel-Aal et al. [27] used a multiscale surface analysis technique to decode the design features of shed skin obtained from *Python regius* (Ball Python). Results pointed at the importance of localized surface design where surface topology, texture, and form vary in accordance with specific functional requirements of different zones along the body. Moreover, the metrological features of the shed skin revealed a multiscale nature of the topographical makeup of the surface, with each scale targeted at optimal function at a particular scale of contact. This customization strongly relates to optimized performance in terms of minimized surface damage and possible economy of energy consumed in combating frictional tractions during locomotion.

The principal idea of artificial texturing is to enhance the sliding performance of a cylinder by imposing a geometrical pattern on the external layer of the liner. That is to create a predetermined geometrical ornamentation on the contacting surface of the cylinder liner. Texturing of the surface results in the creation of a protrusion above the surface, *plateau*, and a channel between any two protrusions, *groove*. Together, the plateau and the groove form a tribological subsystem that helps reduce the destructive effects of sliding between piston and cylinder. The grooves, in principal, are supposed to retain remnants of the lubrication oil during piston sliding and thereby they can replenish the lubrication film in subsequent sliding cycles. The plateau, meanwhile, is supposed to provide raised cushions (islands) that the piston will contact upon sliding, so that reduction of the total contact area takes place. A system of such nature is multiscale by default in the sense that the basic metrological characteristics of a honed surface will have values that depend on the scale of observation. It is this fact that renders the design of a honed surface inspired by the metrological features of a Python surface technically attractive. Multiscale features of conventional textured surfaces achieve optimal performance within a narrow domain of triboconditions, whereas multiscale features of the Python are optimized for a considerably broader range of rubbing situations. As such, if we understand the topology and construction of the Python skin, and their relation to the superior sliding performance of the reptile, then the same design principles may be applied to devise a textured surface of optimized performance within a broad domain of operation conditions.

This chapter details an effort to deduce design rules that allow the construction of bioinspired deterministic surfaces. We draw analogy between the topographical features of the *Python regius* shed skin and those of typical honed surfaces. The scope of the presentation is limited to the metrological analysis of the topographical features rather than delving into the consequence of the surface geometry on frictional response. The presentation proceeds as follows: In the first part, we provide background information about the species under study, its biological features, skin morphology, and essential features of the skin-shedding process. The second section of the chapter details microscopy observations of the shed skin. Further, we report on the metrological aspects of the shed skin topography within the second section. The final part of the chapter presents a comparative analysis between the topographical features of the shed skin and those of typical honed surfaces. Here we identify the differences between natural surface texturing and that implemented through conventional approaches.

4.2 Background

4.2.1 The Python Species

Python regius (Fig. 4.1) is a constrictor type nonvenomous snake species that typically inhabits Africa. An adult snake may become 90–120 cm long. Females tend to be rather larger than males. They may become 120–150 cm long. The build of the snake is nonuniform as the head–neck region, as well as that of the tail, is thinner than the trunk region. Meanwhile, the trunk is the region of the body where most of the snake body mass is concentrated. It is thicker than other parts. The tail section is rather conical in shape (Fig. 4.1a). The overall cross section of the body is more elliptical than circular and the perimeter of the cross section is not uniform along the body. The ventral (stomach) part of the body is typically cream or extremely light yellow in color with occasional black spots scattered within (Fig. 4.1b, d). Skin of the *Python* contains blotches imposed upon an otherwise black background. Shape of the blotches is nonuniform and their colors are dark and light brown (Fig. 4.1c). Form of the body reflects on the contact behavior while sliding. The head will establish minimal contact and so will the tail. The trunk section, however, will be the region where most of the contact interaction takes place. In addition, due to the nonuniform shape, most of the weight of the animal will be concentrated in the trunk section. Consequently, the trunk will constitute the section where most of the volume of the snake establishes contact with the substrate upon sliding. This implies that the skin within the trunk section will accommodate generation of tractions due

4 Reptilian Skin as a Biomimetic Analogue 57

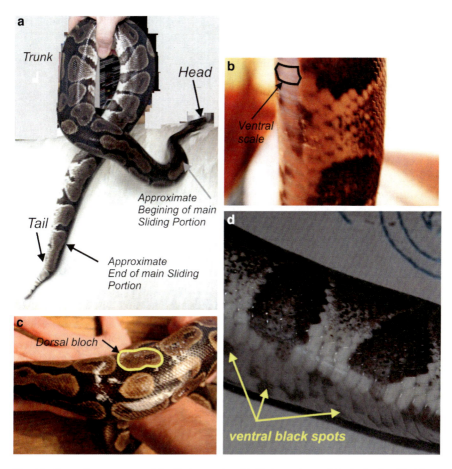

Fig. 4.1 General appearance of the *Python regius*. (**a**) Major regions of the body classified from a locomotion and contact mechanics point of view. Note the nonuniform distribution of the body cross-sectional perimeter andthat the head and tail sections are slender compared to the trunk. (**b**) Ventral side of the snake which is *cream* in color. (**c, d**) Dorsal blotches and ventral black spots

to muscle contractions and also will exhibit the effect of the frictional tractions that resist locomotion.

Scales of various shapes and sizes cover the skin. The scales form by differentiation of the underlying skin (the epidermis). The number, arrangement, size, and shape of the scales vary greatly from one species to another, but are genetically fixed for each species. A snake within a particular species will hatch with a fixed number of scales. The number of scales neither increases nor decreases as the snake matures. Scales, however, grow larger to accommodate growth and may change shape with each molt. Therefore, the patterns of scalation provide simple and accessible recognition characteristics for the taxonomic classification of snake species. The simple body scales overlap one another slightly. Further, a thin interscale layer of skin tissue continuously links them. Snakes periodically molt their scaly skins and acquire new ones.

4.2.2 Structure of Snake Skin

The skin of a snake is a complicated structure. Normally, it contains two generations of skin (see Fig. 4.2 for illustration). The figure shows an *"outer generation layer"* and an *"inner generation layer."* The first represents the layer of skin that is about to be shed. The second represents the layer about to replace the shed skin. Both layers are of the same compositional structure. Like that of many vertebrates, it has two principal layers: the *dermis*, which is the deeper layer of connective tissue with a rich supply of blood vessels and nerves, and the *epidermis*, which in reptiles consists of up to seven sublayers or *"strata"* of closely packed cells, forming the outer protective coating of the body [28]. The "epidermis" has no blood supply, but its inner most living cells obtain their nourishment by the diffusion of substances to and from the capillaries at the surface of the "dermis" directly beneath them.

The epidermis is the layer that directly contacts the surroundings. There are seven epidermal layers as shown in Fig. 4.2. The *"stratum germinativum"* is the deepest layer lining. It contains cells that have the capacity for rapid cell division. Six layers form each *"epidermal generation"* (the old and the new skin layers).

These are the clear layer and the *lacunar layer*, which matures in the old skin layer as the new skin is growing beneath. The *alpha (α)-layer*, the *mesos layer*, and the *beta (β)-layer* consist of cells that get keratinized with the production of two types of keratin (α and β keratin). These cells are thus being transformed into a hard

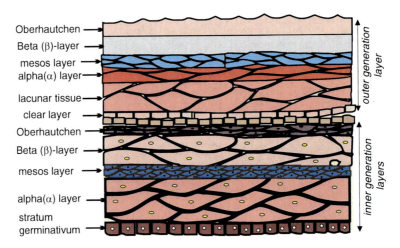

Fig. 4.2 Schematic illustration of a generalized epidermis of a squamate reptile. Two layers are shown: the outer generation layer that represents the skin layer about to be shed and the inner generation layer that represents replacement of the shed layer

protective layer. The final layer is the "*oberhautchen*" that constitutes the strongest outermost layer. It consists of a highly cornified (keratinized) surface, which is covered by a microscopically fine pattern of keels that is a species characteristic. The basale membrane separates the epidermis from the true inner skin (cutis or corium). The cutis consists of a more open and a more solid layer of connective tissue in which there is collagen and elastic fibers.

4.2.3 Skin Shedding

In most mammals, the structure of the epidermis is less complex and the outermost dead skin cells are constantly flaking off; this protective layer constantly replenishes from below. The deepest layer of cells, the "*stratum germinativum*," is constantly dividing and multiplying, so the layers are on the outward move [28]. In reptiles, however, this cell division, in the "*stratum germinativum*," only occurs periodically [29], and when it does, all the layers above it, in the area where the cell division occurs, are replaced entirely. That is, the reptile grows a second skin underneath the old skin, and then "sheds" the old one. About 2 weeks before the reptile sheds its skin, the cells in the *stratum germinativum* begin active growth and a second set of layers forms slowly underneath the old ones. By the end of this time, the reptile effectively has double skin. Following such a process, the cells in the lowest layers of the old skin, the clear and the lacunar layers, and the Oberhautchen layer of the skin below undergo a final maturation and a so-called shedding complex forms. Fluid is exuded and forms a thin liquid layer between them. This gap between the two skins gives a milky appearance to a shedding reptile. Enzymes, in this fluid, break down the connections between the two layers.

The old skin lifts and the reptile actively removes it. This process is shown in Fig. 4.3 that depicts a live species undergoing skin shedding. Note the presence of

Fig. 4.3 Photograph of a *Python regius* during shedding. Two skin layers may be identified: the old (shed) layer almost milky in color and the new replacement that is shiny. Note the flakey appearance of the skin within the magnified zone

two skin layers: the newly generated layer (shinny skin in figure) and the layer that is being shed (milky or opaque appearance).

4.3 Observation of Shed Skin

4.3.1 Initial Observations

Initial observations of the scale structure were performed using photography of a live snake and optical microscopy. All observations took place without the treatment of the skin. Figure 4.4a–e details the surface structure of the live snake. To facilitate the analysis, we defined two virtual axes on the body of the reptile. The first is designated as the major longitudinal body axis. This line coincides with the forward motion of the snake. It is a line centered on the hide of the beast and extends from the tip of the mouth to the tip of the tail (on the dorsal and ventral sides). The second axis is a moving transverse axis. This line coincides with slithering motion. In particular, it coincides with lateral displacements of the body. An illustration of the axes on the head of the animal, on both sides, is provided in Fig. 4.4a, b.

Figure 4.4a–c details the skin geometry in the head region from the inner side (sliding side). A general view of the underside of the head is given in Fig. 4.4a, whereas a general view of the outer side of the head is given in Fig. 4.4b. A close-up of the head–jaw area from the ventral side is shown in Fig. 4.4c. The photographs reveal that polygons constitute the geometrical building block of the surface. This polygon has eight sides, octagon, in the general area of the mouth (represented by the letter *O* in the figure). Past a line that joins the eyes, line AA, the pattern of the skin changes to hexagonal. The hexagonal structure is also dominant within the outer (upper side) of the head as shown in Fig. 4.4b. The size of the hexagonal cells differs from that of the octagonal cells. The size of octagons in the mouth region is not uniform. However, compared with the hexagonal patterns within the throat region, the area of the unit octagon is apparently greater than that of the hexagonal unit. Hexagonal cells on the other hand are of uniform shape and size and seem to be of uniform density per unit area. Within the mouth zone (above the line AA), the aspect ratio of the octagonal unit cell is, qualitatively, uniform. The major axis of the polygon, moreover, appears to point in the same direction of the snake. The uniform distribution of hexagonal cells is likely to aid the compliance of the surface and increase its flexibility. Such a hexagonal pattern is noted to be the most efficient way to pack the largest number of similar objects in a minimum space [30]. Figure 4.4e depicts the surface geometry in the ventral side (general region of the belly). A hexagonal pattern constitutes the basic building block of the skin. The size of the hexagons differs around the circumference of the body. Large cells are particular to the main sliding area, whereas cells of smaller size are particular to the back and the sides. The aspect ratio of the cells is variable. Of interest is the orientation of the major axis of the skin unit cells with respect to the major

4 Reptilian Skin as a Biomimetic Analogue

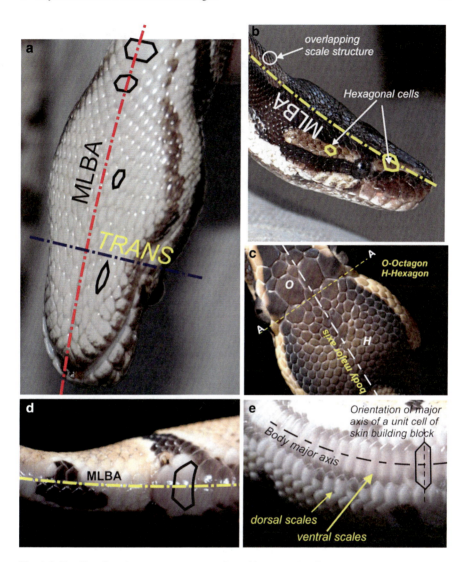

Fig. 4.4 Details of scale structure at several positions on the live snake. (**a**) It presents a generalized view of the throat–jaw region from the sliding side. (**b**) It presents a general view of the outer side of the head. (**c**) It presents a close-up of the jaw–throat area; here the different polygon-based structures of the unit building blocks of the skin are identified. Hexagonal cells are designated H and octagonal cells are labeled O. Note the difference in size between octagonal and hexagonal cells. (**d**) It depicts skin structure within the tail section and (**e**) details the structure within the ventral side of the trunk. Notice that the hexagonal cells have high aspect ratio and that the longest diagonal is oriented perpendicular to the major longitudinal body axis

longitudinal body axis. In the head–throat region, the major axis of the cells is oriented parallel to the body major axis, whereas in the ventral scales, the major cell axis is perpendicular. Varenberg and Gorb [31] based on experiments on the hexagonal structures found on tarsal attachment pads of the bush cricket (*Tettigonia viridissima*) suggest that variation in the aspect ratio of hexagonal structures may alter the friction force of elastomers by at least a factor of two. Additionally, we propose that the perpendicular orientation of the cells, with respect to the major axis of the snake, within the main sliding region aids in shifting the weight, and hence the contact angle and area of the snake upon sliding. Note that since the body of the snake is of cylindrical shape, the highest curvature of the skin will be oriented along the major cell axis. As such, upon sliding, the area of contact, and therefore the total tractions, will depend on the direction of motion (higher sideways and minimal forward). The orientation of the hexagon axis renders the friction forces anisotropic. Such an observation is consistent with the findings of Zhang et al. [29] who studied the frictional mechanism and anisotropy of Burmese Python's ventral scales. They reported that the friction coefficient of the ventral scale had close relationship with moving direction. The frictional coefficient for backward and lateral motion was one-third higher than that in forward motion.

4.3.2 Optical Microscopy Observations

In snakes, the *Oberhautchen* layer is in direct contact with the environment and possesses a fine surface structure called microornamentation [32]. Earlier authors [32–34] described details of the microornamentation. Initial observations on the structure of the scales were performed using optical microscopy without any treatment of the skin. Figure 4.5 depicts the structure of the scales at two positions within the skin in a region close to the waist of the snake. The first was from the back (dorsal scale), whereas the second position represented the stomach of the snake (ventral). Note that although the general form of the cells is quite similar for both positions, the size of a unit cell within the skin is quite different in both cases. In particular, the cell is wider for the ventral positions. Each cell (scale) is also

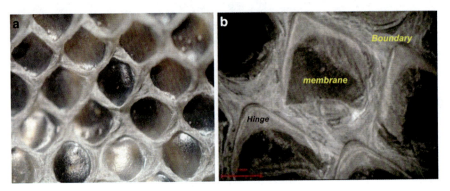

Fig. 4.5 The structure of the scales on the inside of the shed skin at a region close to the midsection of the species at two orientations: back (dorsal) and abdominal (ventral)

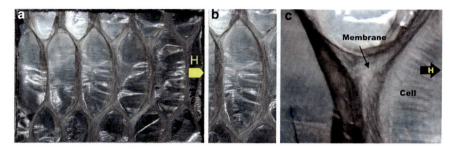

Fig. 4.6 The details of dorsal scales from the inside of shed skin. The terminology used is membrane to denote the major area of the scale and boundary to denote the raised part forming the circumference of the scale

composed of a boundary and a membrane-like structure. Note also the overlapping geometry of the skin and the scales (the so-called scale and hinge structure). The skin from the inner surface hinges back and forms a free area that overlaps the base of the next scale, which emerges below this scale (Fig. 4.5b).

Figure 4.6a–c provides details of the ventral scales. It is noted that the edges of the ventral scales are not straight. Rather, they are curved in the head tail plane with the curvature concave toward the direction of the head (the arrow labeled H in Fig. 4.6a). The overlapping arrangement of the scales and the existence of the elastic connecting tissue are shown in Fig. 4.6b. The figure indicates that the curvature of the leading edge of any one scale (the edge toward the head side) is larger than that of the trailing edge (Fig. 4.6b). A close-up of the hinge region between two cells is depicted in Fig. 4.6c. The hinge is made of the connecting elastic tissue. Note the crisscrossed pattern that distinguishes the elastic tissue area. Note also the size and pattern of the elastic connector tissue (termed as the membrane in Fig. 4.6c) and how it surrounds individual cells.

4.3.3 Scan Electron Microscopy Observations

The shape of the Python, as shown in Fig. 4.1a is essentially nonuniform. The perimeter of the cross sections of the along the body are not equal. In addition, the build of the snake is essentially stocky. As a consequence, the mass of the animal will mainly be concentrated in the middle section of the trunk and the vicinity. Such a form reflects on the sliding behavior of the beast and on the reaction forces that the snake will experience during sliding. To this end, we can consider that the bulk of the friction-induced tractions will affect the region where most of the body mass is concentrated.

A consequence of such a preposition is that the extremes of the trunk region, namely, the neck and the tail regions, will exhibit different friction loading and will undergo different sliding contact mechanics than the bulk of the trunk region.

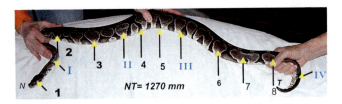

Fig. 4.7 Positions chosen on the shed skin of snake for the observation and characterization of dimensional metrology. Roman numerals indicate principal load-bearing transition regions on the hide of the snake

Table 4.1 Relative position of skin samples chosen for observation and surface parameter evaluation

Position	1	I	2	3	II	4	5	III	6	7	8	IV
X/L	0.01	0.15	0.25	0.45	0.55	0.6	0.65	0.75	0.8	0.85	0.9	0.975

This, in turn, requires, from a functional design point of view, that the local surface geometry and topographical features within the two extremes of the trunk be distinctly different than those of the midsection. In this sense, an analogy between the body of the snake and the surface geometry of a cylinder bore may be developed. Such an attempt draws upon the positions that a piston assumes while sliding during a combustion cycle of an ICE.

In an ICE, a piston assumes three positions. These are the top dead position (TDP), the bottom dead position (BDP), and the mean position (MP). Each of these positions has a distinct lubrication requirement due to the manner that the piston engages the cylinder walls. Further, the local lubrication mechanism operating in each position is different. This, in turn, requires that the local surface texture be compatible with the lubrication regime necessary to maintain proper function and minimize damage. Similar to a snake, the cylinder requires different textural features at the extremes, top, and bottom of the sliding stroke. An analogy is, thus, drawn between the main section of the trunk and the MP. Similarly, the TDP and the BDP are compared to the boundaries of the neck–trunk and the trunk–tail regions, respectively.

To evaluate the parameters of the skin, we identified 12 points on the reptile hide. These are shown in Fig. 4.7. The general location of the selected skin samples are labeled by roman numerals. Table 4.1 presents a summary of the nondimensional position of each area chosen for analysis. The values expressed as x/l entries were obtained by referring the actual distance between the centroid of the respective ventral scale and the nose tip of the reptile to the total length of the snake. Thus, the x/l value represents the distance from the nose point, N, to the centroid of the particular ventral scale divided by the distance from the nose to the tail point, which is roughly about 11,270 mm. The roman numerals in the figure denote key locations on the hide from which skin swatches were selected for SEM observations. The points denoted I and IV identify ventral scales located at the top and bottom boundaries of the trunk section, respectively. Thus, point I identifies a ventral scale roughly located within the TDP region and point IV identifies a scale located within the BDP region. Points II and III meanwhile refer to scales located on the front half of the skin and the rear half of the skin, respectively. Both scales, however, are located within the MP region.

Skin swatches from each of the chosen positions were examined at different magnifications ($X = 250 - X = 15,000$) in topography mode. In order to suppress charging phenomena and improve the quality of observation, the surface of each sample was coated with a 10-nm-thick layer of platinum (Pt) using a sputter coater (EMITECH K575X).

For each position, samples from the dark- and the light-colored skin (see Fig. 4.5) were also examined along with samples from the underside of the body. Major features of the observations are shown in Figs. 4.8 and 4.9.

Figure 4.8a–e depicts SEM micrographs taken at a magnification of $X = 250$, whereas Fig. 4.9a–e provides the micrographs of the same skin swatches depicted in Fig. 4.8 at a magnification of $X = 250$. Thus, the pictures provided in Fig. 4.9 are at one-to-one correspondence with those provided in Fig. 4.8. The scale marker in Fig. 4.8 is $100\,\mu m$ and that in Fig. 4.9 is $5\,\mu m$. Photographs of two positions are depicted in the figures. In both figures, the left-hand side columns (labeled a, c, and e) depict photographs of a zone located within the center of the boundary of the cell (i.e., within the elastic connector tissue between two cells). The pictures labeled (b, d, and f), within both figures, detail zones within the scale membrane. From top to bottom, within each figure, the photographs labeled a and b depict details of the ventral scales, those labeled c and d depict scales within the bright (light) dorsal skin, and finally, those labeled e and f provide details of scales within the dark skin.

Figure 4.8 reveals a grainy appearance of the scale boundary irrespective of the side of the body (dorsal or ventral). The membrane structure, on the other hand, reveals a wavy appearance at the low magnification. This grainy appearance, within the scale boundary, manifests microprotrusions that appear to be of random shape and distribution. Size and volume of these protrusions appear to be quasi-uniform. The protrusions at this scale appear to be manifest folds within the elastic tissue. The wavy appearance of the membrane, however, appears to be in an overlapping arrangement. The spacing between waves seems to be more compact on the dorsal scales (Fig. 4.8d, f) than the spacing within the ventral scales (Fig. 4.8b).

Figure 4.9 reveals that the protrusions within the scale boundaries are approximately hemispherical (or at least concave clusters) and comprise pores. The space between the waves in the ventral scales, as well as the located dorsal scales, is also full of pores. Dorsal scales located within the dark skin do not appear to contain pores. Two types of pores (or micropits) may be distinguished: those located within the boundary and those located within the membrane. Image analysis of the pictures indicates that the diameter of the boundary-pores ranges between 200 and 250 nm. The diameter of the membrane-pores was estimated by Hazel et al. [13] using AFM analysis to be in the range of 50–75 nm.

The surface of the membrane also comprises micro-nano-fibrile structures. These are not of consistent shape and spacing. Note, for example, that the shape of fibril located in the dark-colored skin region is different than that located within the light-colored skin region (compare the X-5,000 pictures). The fibrils within the

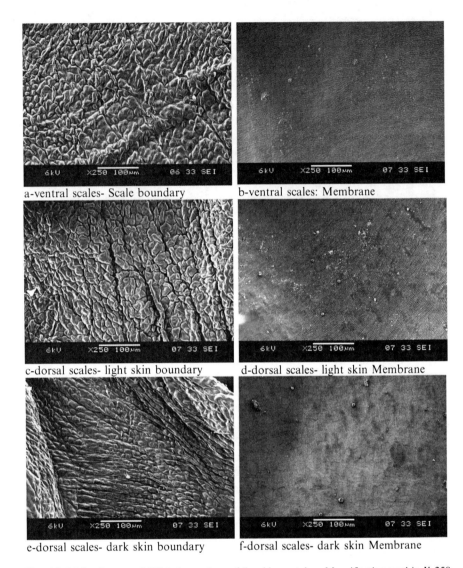

Fig. 4.8 Major features of SEM observations of the skin swatches. Magnification used is X-250 and the *scale bar* is 100 μm. (**a**) Ventral scales – scale boundary. (**b**) Ventral scales – membrane. (**c**) Dorsal scales – light-skin boundary. (**d**) Dorsal scales – light-skin membrane. (**e**) Dorsal scales – dark-skin boundary. (**f**) Dorsal scales – dark-skin membrane

dark-colored scales are longer than those located within the dorsal bright skin and within the ventral scales. The width of the fibrils, in both regions, appears to be different. Fibril tips are pointed toward the tail. Within the dark dorsal scales, fibrils are tapered and have a sharp tip. Scales within the bright-colored and the ventral regions have a more rounded tip and appear to be of uniform width throughout the

4 Reptilian Skin as a Biomimetic Analogue

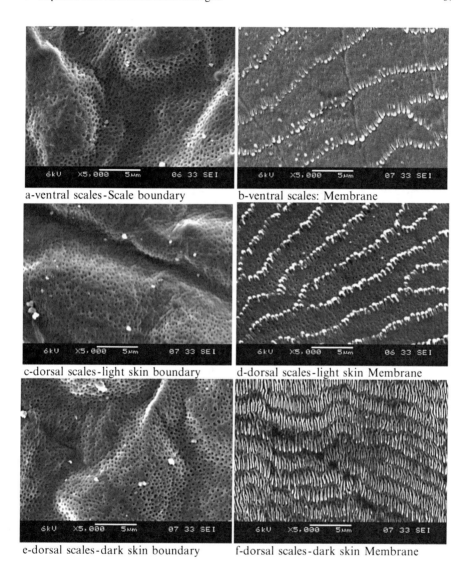

Fig. 4.9 Major features of SEM observations of the skin swatches at a magnification of $X = 5,000$. The *scale marker* is $5\,\mu m$. Pictures are in one-to-one correspondence with those provided in Fig. 4.10. (**a**) Ventral scales – scale boundary. (**b**) Ventral scales – membrane. (**c**) Dorsal scales – light-skin boundary. (**d**) Dorsal scales – light skin membrane. (**e**) Dorsal scales – dark skin boundary. (**f**) Dorsal scales – dark skin membrane

fibril length. Moreover, the density of the fibrils seems to be different within the different color regions (denser within the dark-colored region).

For each point shown in Fig. 4.7, a series of five SEM pictures at different locations within the particular ventral scale was recorded. The pictures were further

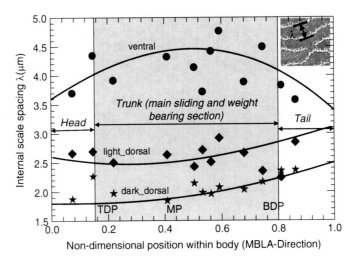

Fig. 4.10 Variation in the intraspacing between rows of the microfibrils located within the membrane (hinge) of the snake scales as a function of: distance along the MBLA, position of scales, and color

analyzed to obtain fibril geometric information (counts, distance between fibrils, and length of individual fibrils). The average (arithmetic mean) of the information sets of the selected ventral scales were then plotted against the nondimensional distance. Here we present the variation in the distance between waves of fibrils. To verify this observation, we measured the distance between rows of fibrils (wave spacing) from SEM micrographs taken at each of the positions depicted in Fig. 4.7. In all, 12 positions were examined.

Figure 4.10 presents a plot of the internal spacing, λ, (distance between fibril rows in micrometer) as a function of the nondimensional distance X/L. Location on the skin may be obtained by comparing the X/L values to entries in Table 4.1. Internal spacings in the figure represent the average of five separate measurements within different regions of the same SEM picture. As such, each data point in the figure is actually an average of 25 readings on the individual ventral scale. Approximate boundaries of the trunk are located within the shaded rectangle in the plot. The markings TDP, MP, and BDP refer to the same regions identified in Fig. 4.10.

Data from the figure indicate that the distance between fibril waves vary between $3.5 < l < .4.8\,\mu$m in the ventral scales. The variation of that distance within the dorsal scales is $2 < \lambda < 3\,\mu$m and $1.5 < \lambda < 2.4\,\mu$m for the light-colored and dark-colored dorsal skin, respectively. The shorter spacing is roughly located within the non-load-bearing portions of the body (i.e., the head and tail sections). The distribution of the separation distance λ along the body is not uniform. The internal spacing is larger within the general region of the trunk. The maximum spacing is roughly located within the middle section of the trunk (MP-S). It is to be remembered that the arrangement of the fibrils does not constitute straight

4 Reptilian Skin as a Biomimetic Analogue

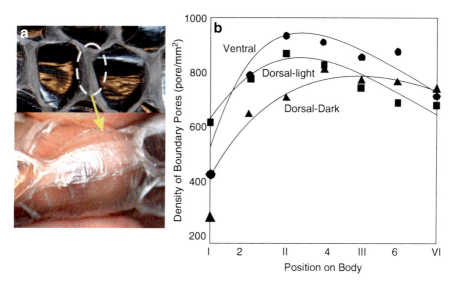

Fig. 4.11 The variation in the density of the boundary pores (pore/mm²) with position and with color of skin

lines. Note that the internal spacing between different rows of fibrils also differs by skin color and position within the body in the order $\lambda_{ventral} > \lambda_{dorsal\ light\ skin} > \lambda_{dorsal\ dark\ skin}$.

Further analysis of images revealed that the density of the boundary pores vary by position. The number of pores per unit area is not constant along the body. It changes relative to the position within the skin. Figure 4.11 is a plot of the variation in the density of the pores relative to the two sides of the skin (back-dorsal scales and abdominal-ventral scales) and in relation to the color of the skin (light patches vs dark patches) within the back also.

Density of pores was obtained from SEM micrographs by counting the pores in the picture and dividing by the area of the region scanned by the picture. Each point in the graph is an average of five counts taken from five different SEM observations of regions located within the same general area of analysis. Consistent with the trend noted in Fig. 4.10, the plot indicates that the pore density is higher in the order ventral > dorsal light > dorsal dark.

4.4 Metrology of the Surface

4.4.1 Topographical Metrology

To characterize the surface topography of the skin, we selected several swatches of skin (1,500 μm × 1,500 μm) for examination using white light interferometry (WLI). The results yielded the basic parameters that describe the surface (asperity

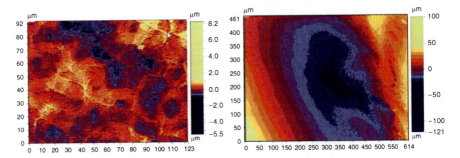

Fig. 4.12 Multiscale WLI graphs depicting the topography of the skin building block (scale) boundary and membrane

radii and curvature etc.). Figure 4.12 depicts a typical WLI graph of the skin. The shown inteferogram pertains to a skin spot that is located along the waist of the snake from the belly side (ventral). Two interferograms are depicted: the one to the right-hand side of the figure represents the topography of the cell membrane, whereas the one depicted to the left represents a multiscale scan for the whole skin swatch. Note the scale on the right of the pictures as it indicates the deepest valley and highest point of the skin topography. For these skin swatches, the value of the deepest part of the membrane was about 120 μm, whereas the highest summit is about 100 μm. The comparable values for the whole swatch are about 5.5 and 8.2 μm, respectively. These initial observations prompted the study of the so-called Abbott–Firestone load-bearing curve at different locations within the skin.

4.4.2 Bearing Curve Analysis

Surfaces, irrespective of their method of formation, contain irregularities or deviations from a prescribed geometrical form. The high points on the surfaces are referred to as asperities, peaks, summits, or hills and the low points as valleys. When two rough surfaces acted upon by a normal force come into contact, the opposite surface peaks to make contact first, are those for which the sum of the heights is the longest. As the load is increased, new pairs of opposite peaks having an even smaller sum of heights will be coming in contact. Once in contact, the surface peaks become deformed. This deformation leads to an increase in the contour area of contact, and as a result, to an increase in the number of peaks sustaining the load. Since the peaks differ in height, the deformation of various peaks on one hand the same surface will be different at any instant of time. The irregularities of the mating surfaces are out of contact over a considerable portion of the apparent contact area because of surface waviness and form errors. However, only of a fraction of the apparent area establishing the contact will actually bear a load. A question that arises in such a situation is: given a known surface roughness profile, how can we calculate the area that actually supports a load? The answer is normally formulated in terms of a so-called load bearing area curve (also known as the *Abbott–Firestone load curve* (AFLC)). The idea of that curve is to calculate the probability of a roughness protrusion to of a given height establishes true contact with a virtually smooth surface. Computing such a probability for a series of heights that sufficiently describe the surface, from a statistical point of view, yields a probability density distribution that relates to the true profile of the surface. Upon integrating this distribution with respect to surface height, we obtain the AFLC [35–37]. Study of the AFLC yields a worthy prospective of the potential behavior of a given surface upon sliding and the potential for damage through wear. Figure 4.13 presents the AFLC for three ventral scales on the live snake. These are highlighted in the

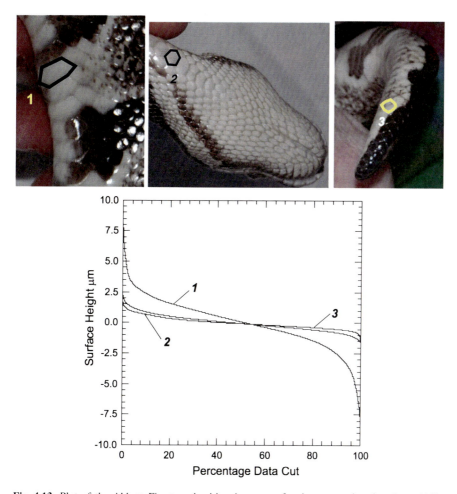

Fig. 4.13 Plot of the Abbott–Firestone load-bearing curve for three ventral scales, 1 – middle section of the trunk, 2 – neck-trunk boundary, and 3 – tail section

photographs labeled 1, 2, and 3 in the figure. The picture labeled 1 depicts a ventral scale located in the middle section of the trunk (MP region). The picture labeled 2 depicts a ventral scale located within the general area of the throat-neck (TDP region). Finally, the photograph labeled 3 depicts a ventral scale located within the tail section. Plots in the figure are labeled accordingly.

The plot reflects the symmetry of the ventral scales. The individual plots are rotationally symmetric around the vertical axis (percentage data cut). On the other hand, there is a significant difference in the load-bearing capacity of the middle section, position 1, and the other two regions (2 and 3). This is interesting as it supports the customization in natural design. As mentioned earlier, position 1 represents the zone where the bulk of the mass of the snake is contained. It also represents the region, within the ventral side of the snake, where most of the frictional tractions generated while locomotion is likely to be concentrated. In contrast to this region, zones 2 and 3 are not likely to be as loaded. Consequently, the probability of sustaining severe damage in sliding is not prominent. From a design point of view, there is no requirement to enforce the thickness of the surface. It is also apparent from the figure that different zones within the skin have variable surface profile parameters that are specific to their frictional profile.

Further analysis shows that the AFLC yields predictive information about sliding performance. The predictions formulate standardized surface-functionality assessment parameters. Of interest in this presentation is the so-called R_k family of parameters [38–40] and their equivalent CNOMO counterparts [41].

We studied the load-bearing characteristics of the skin at each of the key positions (I through IV). Surface parameters were extracted from SEM topography photographs. The complete set of analyzed pictures provided a matrix of roughness parameters that describe the texture of the shed skin at variable scales ranging from X-100 to X-5,000. Table 4.2a, b provides a summary of the parameters extracted from the analysis. It can be seen that the scale of the analysis affects the value of the parameters, which may point at a fractal nature of the surface.

Comparing the ratios between the reduced peak height R_{pk}, core roughness depth R_k and reduced valley depth R_{vk} reveals symmetry between the positions (compare

Table 4.2 Effect of magnification on surface parameters as deduced from SEM micrographs

	C_r/C_f	C_l/C_f	R_{pk}/R_k	R_{vk}/R_k	R_{vk}/R_{pk}
(a) Surface parameters based on X-250 pictures					
Position I	0.718	0.861	0.391	0.159	1.144
Position II	2.011	2.010	0.612	0.545	0.656
Position III	2.066	1.628	0.733	0.436	0.621
Position VI	1.388	0.926	0.617	0.195	0.749
(b) Surface parameters based on X-5,000 pictures					
Position I	1.930	1.207	0.654	0.285	0.679
Position II	1.273	1.803	0.478	0.404	0.812
Position III	1.671	1.622	0.484	0.359	0.800
Position VI	1.772	1.158	0.636	0.260	0.688

4 Reptilian Skin as a Biomimetic Analogue

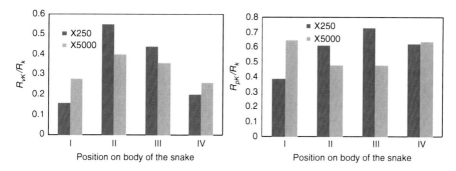

Fig. 4.14 Plot of the ratio of the load-bearing parameters R_{vk}/R_k and R_{pk}/R_k at two magnifications X-250 and X-5,000

the columns R_{pk}/R_k, R_{vk}/R_k, and R_{vk}/R_{pk} of Table 4.2b, and Fig. 4.14a, b). This symmetry is interesting on the count that positions II and III represent the boundaries of the main load-bearing regions (trunk). This is the region within the body where the snake has most of its body weight concentrated (refer to Fig. 4.5). Thereby, it is the region that is principally used in locomotion. The symmetry in surface design ratios is more apparent at higher magnification (X-5000). This implies that the symmetry is more significant at small scale, which points out at the uniformity of the surface basic building blocks at smaller sizes. Such symmetry may very well be related to the wear resistance ability of the surface or to the boundary lubrication quality of locomotion. It is of interest to find whether implementing a surface of such characteristic parameters (functionally textured surface) in plateau honing, for example, would be conducive to an antiscuffing and economical lubricant consumption performance.

The motivation for such a proposal stems from identification of the common, functional and geometrical, features between the surface of the Python and that of a honed surface.

4.5 Correlation to Honed Surfaces

The principal idea of honing is to enhance the wear and sliding performance of the cylinder liner by imposing a geometrical pattern on the external layer of the liner, that is, to create a predetermined geometrical (artificial ornamentation) on the contacting surface of the cylinder liner.

The texturing of the surface has two basic components: a raised protrusion called "*plateau*" and an entrenched component known as a "*groove*." Together, the plateau and the groove form a tribological subsystem that aids in reducing the destructive effects of sliding between piston and cylinder. The grooves, in principal, are supposed to retain remnants of the lubrication oil during piston sliding and thereby they can replenish the lubrication film in subsequent sliding cycles. The plateau, meanwhile, is supposed to provide raised cushions (islands) that the piston will contact upon sliding, so that the total contact area is reduced. A system of such nature is multiscale by default. That is, the basic metrological characteristics of a honed surface will have values that depend on the scale of observation. Referring to Figs. 4.2 and 4.3, one would notice that on a large scale, that of human eye observation, the texture of the skin of a Python resembles that of a honed surface. Indeed, one observes the existence of *"plateaux"* (hexagonal dorsal and ventral scales) and grooves (boundaries between scales). Knowing the superb tribological performance of the Python, a comparison of metrological characteristics between skin and a honed surface seems in order.

One of the major requirements for an optimal honed surface is connectedness. Through perfect connectedness between all the surface unit-texture features, high lubrication quality, and economical lubricant consumption, is supposed to take place. Economical oil consumption, in essence, takes place because of controlled bearing curve surface features (e.g., R_{pk} and R_{vk}). To date, there are no standardized values of relevant surface parameters to ensure superior performance of a honed cylinder liner. Instead, each manufacturer has in-house set of ranges that surface parameters are supposed to fall within for quality performance. To this end, a comparison between the basic metrological features of a Python skin and those of, medium-quality surface (say) would highlight, at least qualitatively, an optimal range of the geometrical proportions should be maintained within a high-quality performing surface.

To compare skin and honed surface, segments of a commercial unused engine prepared by a sequence of honing processes were examined. Details of surface preparation sequence of operations and properties of the material are summarized in Table 4.3. Figure 4.15a, b depicts two SEM micrographs of one of the segments used for comparison with skin at two magnifications (X-250 and X-5,000). All pictures were obtained after cleaning the surfaces by acetone in an ultrasonic bath. To obtain surface geometry parameters, each SEM micrograph was analyzed using the same program that is used to analyze the skin samples.

Results of the analysis are given in Fig. 4.16a, b compared to those extracted from the skin of the Python. The figure depicts metrological surface parameters calculated at two scales of observation: X-250 and X-5,000. The parameters designated Python were obtained by averaging the values of the particular ratio of the zones II and III on the Python skin (refer to Fig. 4.5 and the values presented in Fig. 4.16a, b along with Table 4.2). The plateau-honed surface (designated as PH in figure) is classified as a so-called grade three surface [41]. According to the standard adopted in this case, such a surface, while not a superior finished surface, is still acceptable from a quality control point of view. As such the comparison made in Fig. 4.16 is between the geometrical surface proportions of the main sliding zone in the Python (zones II and III) and the parameters of an average quality surface that is applied in real practice.

4 Reptilian Skin as a Biomimetic Analogue

Table 4.3 Surface parameters for different regions within the skin as deduced from WLI interferograms

Position within skin	R_k (nm)	R_{pk} (nm)	R_{vk} (nm)	R_{pk}/R_k	R_{vk}/R_k	R_{vk}/R_{pk}
Bearing curve parameters						
Ventral outside (exterior)	975.08	705.3	438.55	0.723	0.449	0.622
Close to tail inside	1,126.53	409.91	419.8	0.364	0.373	1.024
Close to waist inside	3,671.12	2,690.06	4,101.7	0.733	1.117	1.525
Close to waist outside	4,757	2,458.51	3,197.23	0.517	0.672	1.30
Major surface roughness variables						
Position within skin	R_a (nm)	R_q (nm)				
Ventral inside	357.96	477.89				
Close to tail inside	337.04	426.35				
Close to waist inside (dorsal)	1,50	2,790				
Close to waist outside (dorsal)	1,590	2,140				

Fig. 4.15 SEM micrographs of a cylinder liner sample used for comparison with the Python skin, X-250 – *left-hand side picture* and X-5,000 – *right-hand side picture*

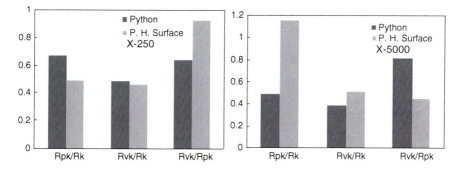

Fig. 4.16 Comparison between the geometrical and metrological proportions of the Python skin surface and those of a plateau-honed cylinder liner

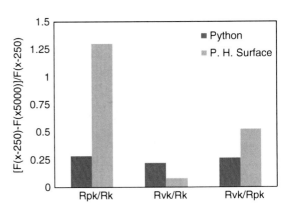

Fig. 4.17 Absolute value of percentage variations in surface proportions as a function of scale of observation

To establish a *qualitative* reference for analysis, we recall the relation of each of the surface parameters R_{pk}, R_{vk}, and R_k to surface performance in sliding. R_{pk} is related to the amount of the surface that will be worn away during the run-in period, which is preferred to be at a minimum. R_k relates to the working portion of the surface that will carry the load after the run-in period, which is preferred to have a relatively higher value. R_{vk} meanwhile relates to the lowest portion of the surface that will retain lubricant, which for minimum oil consumption is required to be relatively high. As such, again qualitatively, the ratio R_{pk}/R_k should be in the neighborhood of 0.5 or less, R_{vk}/R_k should be relatively smaller than unity at or around 0.5, and R_{vk}/R_{pk} should be relatively high at or around unity or slightly higher. Again we emphasize that such limits are qualitative estimates for the sake of comparison. Thus, proceeding from these estimates, one would notice that the Python skin is slightly off the target limits at large scale (X-250). An opposite trend, however, is noted on smaller scale (X-5,000). Interestingly, however, while the ratio R_{pk}/R_k is almost within qualitative limits for the honed surface at large scale (X-250), it significantly departs from that limit at smaller scale (X-5,000). The opposite is noted for the ratio R_{vk}/R_{pk}. Such an observation prompted the calculation of the percentage of change in the three examined geometrical proportions as a function of scale of observation. Results are plotted in Fig. 4.17.

It has been observed that the Python skin has uniform, and minimal, variation in the surface geometrical proportions than the honed surface. For the honed surface, the ratio R_{pk}/R_k shows the largest variation. Such a wide variation may affect performance of the honed surface at very small scale by affecting the connectedness of the microgrooves and thereby the unobstructed flow of the lubricant while sliding. Observation scale is shown in Fig. 4.17. There are many similarities between the honed surfaces and the geometry of Pythons and snakes, in general, that warrant additional studies.

The geometry of a honed surface, its texture and ornamentation, is introduced through the action of the honing tool. The pattern comprising the surface texture, similar to that of the skin of a Python, is composed of small-size building blocks. In a honed surface, the basic surface unit is a four-sided polygon, a parallelogram, whereas in the Python skin, it is mainly a hexagon that differs in the aspect ratio according to location within the body. Similar to a Python, the parallelograms in a honed surface protrude above the main surface. For optimal performance, a major requirement for honed surfaces is perfect connectedness of between all the grooves separating the unit surface texture building blocks. Perfect connectedness ensures unobstructed oil flow for lubrication of sliding surfaces and retention of oil to replenish the surface with lubricant to ensure separation of surfaces upon contact. Perfect connectedness is a major feature of the Python skin as was observed in Figs. 4.4–4.6. Connectedness is also ensured at a smaller scale that renders the Python skin more appropriate than that of the honed surface examined. An additional criterion for an optimal honed surface is the absence of wear debris, that are of the same or larger width than the groove, within the grooves. For Pythons, and snakes in general, maintaining skin cleanliness depends on preventing debris from the environment, or dust particles from clogging lipid passages within the body. The mechanism responsible for this function in a snake depends on the consistency of geometry and metrology of the skin surface. Such consistency is evident from the trend of change in proportions of the main surface parameters as a function of magnification.

4.6 Conclusions and Future Outlook

In this chapter, we presented the results of an initial study to probe the geometric features of the skin of the *Python regius*. It was found that the structure of the unit cells is of regionally similar shape (octagonal and hexagonal).

Although almost identical in size and density, the skin constituents (pore density and essential size of the unit cell) vary by position on the body. Analysis of the surface roughness parameters implied a multiscale dependency of the parameters. This may point at a fractal nature of the surface a proposition that needs future verification.

The analysis of bearing curve characteristics revealed symmetry between the front and back sections of the snake body. It also revealed that the trunk region is bounded by two cross sections of identical bearing curve ratios. This has implications in design of textured surfaces that retain an unbreakable boundary lubrication quality and high wear resistance.

Clearly, much work is needed to further probe the essential features of the surface geometry, namely, the basic parametric make up of the topography and its relation to friction and wear resistance.

References

1. R. Van Basshysen, F. Schfer (eds.), Internal Combustion Engine Handbook, Basics, Components, Systems and Perspectives (SAE International Warrendale, PA, Chapter 9, 2004)
2. M. Priest, C.M. Taylor, Automobile engine tribology-approaching the surface. Wear **241**, 193–203 (2000).
3. E. Willis, Surface finish in relation to cylinder liners, Wear **109** (1–4), 351–366 (1986)
4. N.W. Bolander, F. Sadeghi, Deterministic modeling of honed cylinder liner friction. Tribol. Trans. **50** (2), 248–256 (2007)
5. M. Santochi, M. Vignale, A study on the functional properties of a honed surface. Ann. CIRP **31**, 431–434 (1982)
6. I. Etsion, Y. Kligerman, G. Halperin, Analytical and experimental investigation of laser-textured mechanical seal faces. Tribol. Trans. **42**, 511–516 (1999)
7. I. Etsion, G. Halperin, A laser surface textured hydrostatic mechanical seal. Tribol. Trans. **45**, 430–434 (2002)
8. G. Ryk, Y. Kligerman, I. Etsion, Experimental investigation of laser surface texturing for reciprocating automotive components. Tribol. Trans. **45**, 444–449 (2002)
9. A. Ronen, I. Etsion, Y. Kligerman, Friction-reducing surface texturing in reciprocating automotive components. Tribol. Trans. **44**, 359–366 (2001)
10. L.F. Toth, What the bees know and what they don't know. Bull. Amer. Math. Soc. 70, 4, 468–481 (1964)
11. E. Arzt, S. Gorb, R. Spolenak, From micro to nano contacts in biological attachment devices. *Proc. Natl. Acad. Sci.* 100 (19), 10603 (2003)
12. I. Rechenberg, Tribological characteristics of sandfish, in *Nature as Engineer and Teacher: Learning for Technology from Biological Systems* (Shanghai, Oct. 8–11, 2003)
13. J. Hazel, M. Stoneb, M.S. Gracec, V.V. Tsukruk, Nanoscale design of snake skin for reptation locomotions via friction anisotropy. J. Biomech. 32, 5, 477–484 (1999)
14. L.J. Vitt, E.R. Pianka, W.E. Cooper Jr., K. Schwenk, History and the *global ecology* of *squamate reptiles*. Am. Nat. 162, 44–60 (2003)
15. C. Chang, et al. Reptile scale paradigm: Evo-Devo, pattern formation and regeneration. Int. J. Dev. Biol. 53, 813–826 (2009)
16. L. Alibardi, M.B. Thompson, Keratinization and ultrastructure of the epidermis of late embryonic stages in the alligator (*Alligator mississippiensis*). J. Anat. 201, 71–84 (2002)
17. R. Ruibal, The ultrastructure of the surface of lizard scales. Copeia (4): 698–703 (1968)
18. R.B. Chiasson, D.L. Bentley, Lowe CH scale morphology in *Agkistrodon* and closely related crotaline genera. Herpetologica 45, 430–438 (1989)
19. B.C. Jayne, Mechanical behaviour of snake skin. J. Zool., London 214, 125–140 (1988)
20. M. Scherge, S.N. Gorb, *Biological Micro- and Nanotribology* (Springer-Verlag, Berlin Heidelberg, 2001)
21. G. Rivera, A.H. Savitzky, J.A. Hinkley, Mechanical properties of the integument of the common gartersnake, *Thamnophis sirtalis* (Serpentes: Colubridae). J. Exp. Biol. 208, 2913–2922 (2005)
22. R.A. Berthé, G. Westhoff, H. Bleckmann, S.N. Gorb, Surface structure and frictional properties of the skin of the Amazon tree boa *Corallus hortulanus* (Squamata, Boidae). J. Comp. Physiol. A Neuroethol. Sens. Neural Behav. Physiol. 195(3), 311–318 (2009)
23. M. Saito, M. Fukaya, T. Iwasaki, Serpentine locomotion with *robotics snakes*. IEEE Control Syst. Mag. 20 (2), 64–81 (2000)
24. A.B. Safer, M.S. Grace, Infrared imaging in vipers: differential responses of crotaline and viperine snakes to paired thermal targets. Behav. Brain Res. 154, 1, 55–61 (2004)
25. M. Shafiei, A.T. Alpas, Fabrication of biotextured nanocrystalline nickel films for the reduction and control of friction. Mater. Sci. Engin.: C, 28, 1340–1346 (2008)
26. M. Shafiei, A.T. Alpas, Nanocrystalline nickel films with lotus leaf texture for superhydrophobic and low friction surfaces. Appl. Surf. Sci. 256, 3, 710–719 (2009)

27. H.A. Abdel-Aal, M. El Mansori, S. Mezghani, Multi-scale investigation of surface topography of Ball Python (*Python regius*) shed skin in comparison to human skin. Trib. Lett. 37, 3, 517–528 (2010), DOI. 10.1007/s11249-009-9547-y
28. P. Wu, et al. Evo-Devo of amniote integuments and appendages. Int. J. Dev. Biol. 48, 249–270 (2004)
29. H. Zhang, et al. Structure and friction characteristics of snake abdomen. Nanjing Hangkong Hangtian Daxue Xuebao/J. Nanjing Univ. Aeronaut. Astronaut. 40(3), 360–363 (2008)
30. P. Ball, *The Self-Made Tapestry: Pattern Formation in Nature* (Oxford University Press, New York, 2001)
31. M. Varenberg, S.N. Gorb, Hexagonal surface micropattern for dry and wet friction. Adv. Mater. 21(4), 483–486 (2009)
32. R.B. Chiasson, D.L. Bentley, Lowe CH scale morphology in *Agkistrodon* and closely related crotaline genera. Herpetologica 45, 430–438 (1989)
33. D.J. Gower, Scale microornamentation of uropeltid snakes. J. Morphol. 258, 249–268 (2003)
34. C. Chang, et al., Reptile scale paradigm: Evo-Devo, pattern formation and regeneration. Int. J. Dev. Biol. 53, 813–826 (2009)
35. E.J. Abbott, F.A. Firestone, Specifying surface quality: a method based on accurate measurement and comparison. Mech. Engin. 55, 569–572 (1933)
36. T. Tsukizoe, *Precision Metrology* (Yokkendo Publishing, Tokyo, Japan, 1970)
37. J.B.P. Williamson, The shape of surfaces, in *CRC Handbook of Lubrication*, vol. II, ed. by E.R. Booser (CRC Press, Inc., Boca Raton, FL, 1984)
38. ISO 13565-2, 1996. ISO 13565-2 (1st ed.), Geometrical product specifications (GPS) – Surface texture: Profile method; Surface having stratified functional properties – Part 2: Height characterization using linear material ratio, International Organization for Standardization, Geneva, Switzerland (1996)
39. ISO 13565-3, 1998. ISO 13565-3 (1st ed.), Geometrical product specifications (GPS) – Surface texture: Profile method; Surface having stratified functional properties – Part 3: Height characterization using material probability curve, International Organization for Standardization, Geneva, Switzerland (1998)
40. GE40–087G, Guides des Biens d'Equipement, PSA PEUGEOT – CITROËN, 2007
41. Etats et aspects de surface des chemises et futs de carter-cylindres apres un pierrage plateau, Normalisation Renault Automobiles, DMC/Service 65810:34–09–929/-E, 1997

Chapter 5
Multiscale Homogenization Theory: An Analysis Tool for Revealing Mechanical Design Principles in Bone and Bone Replacement Materials

Christian Hellmich, Andreas Fritsch, and Luc Dormieux

Abstract Biomimetics deals with the application of nature-made "design solutions" to the realm of engineering. In the quest to understand mechanical implications of structural hierarchies found in biological materials, multiscale mechanics may hold the key to understand "building plans" inherent to entire material classes, here bone and bone replacement materials. Analyzing a multitude of biophysical hierarchical and biomechanical experiments through homogenization theories for upscaling stiffness and strength properties reveals the following design principles: The elementary component "collagen" induces, right at the nanolevel, the mechanical anisotropy of bone materials, which is amplified by fibrillar collagen-based structures at the 100-nm scale, and by pores in the micrometer-to-millimeter regime. Hydroxyapatite minerals are poorly organized, and provide stiffness and strength in a quasi-brittle manner. Water layers between hydroxyapatite crystals govern the inelastic behavior of the nanocomposite, unless the "collagen reinforcement" breaks. Bone replacement materials should mimic these "microstructural mechanics" features as closely as possible if an imitation of the natural form of bone is desired (Gebeshuber et al., Adv Mater Res 74:265–268, 2009).

Nomenclature

$\mathbb{D}_{rs}$ Fourth-order influence tensor
$\mathbb{A}_r$ Fourth-order strain concentration tensor of phase r
$\mathbb{C}_{\text{col}}$ Fourth-order stiffness tensor of molecular collagen
$c_{\text{col},ijkl}$ Component of fourth-order stiffness tensor of molecular collagen

---

C. Hellmich (✉)
Institute for Mechanics of Materials and Structures, Vienna University of Technology (TU Wien), Karlsplatz 13/202, 1040 Vienna, Austria
e-mail: christian.hellmich@tuwien.ac.at

$\mathbb{C}_r$	Fourth-order stiffness tensor of phase r
$\mathbb{C}^{hom}$	Homogenized fourth-order stiffness tensor
$\mathbb{C}^0$	Fourth-order stiffness tensor of an infinite matrix surrounding an ellipsoidal inclusion
d	Characteristic length of the inhomogeneities within an RVE
E	Second-order "macroscopic" strain tensor
E^p	Second-order "macroscopic" plastic strain tensor
$\underline{e}_1, \underline{e}_2, \underline{e}_3$	Unit base vectors of Cartesian reference base frame
$\underline{e}_\vartheta, \underline{e}_\varphi, \underline{e}_r$	Unit base vectors of Cartesian local base frame of a single crystal of hydroxyapatite within extrafibrillar space
$f_r(\sigma_r)$	Boundary r of elastic domain of phase r in space of microstresses
$\bar{f}_{col}$	Volume fraction of collagen within an RVE $\bar{V}_{excel}$
$\mathring{f}_{col}$	Volume fraction of molecular collagen within an RVE $\mathring{V}_{wetcol}$
$\bar{f}_{ef}$	Volume fraction of extrafibrillar space within an RVE $\bar{V}_{excel}$
$\tilde{f}_{excel}$	Volume fraction of extracellular bone matrix within an RVE $\tilde{V}_{exvas}$
f_{exvas}	Volume fraction of extravascular bone material within an RVE V_{cort}
$\bar{f}_{fib}$	Volume fraction of mineralized collagen fibril within an RVE $\bar{V}_{excel}$
$\bar{f}_{HA}$	Volume fraction of hydroxyapatite within an RVE $\bar{V}_{excel}$
$\check{f}_{HA}$	Volume fraction of hydroxyapatite within an RVE $\check{V}_{fib}$
$\check{f}_{HA}$	Volume fraction of hydroxyapatite within an RVE $\check{V}_{ef}$
$\check{f}_{ic}$	Volume fraction of intercrystalline space within an RVE $\check{V}_{ef}$
$\mathring{f}_{im}$	Volume fraction of intermolecular water within an RVE $\mathring{V}_{wetcol}$
$\tilde{f}_{lac}$	Volume fraction of lacunae within an RVE $\tilde{V}_{exvas}$
f_r	Volume fraction of phase r
f_{vas}	Volume fraction of Haversian canals within an RVE V_{cort}
$\tilde{f}_{wetcol}$	Volume fraction of wet collagen within an RVE $\check{V}_{fib}$
HA	Hydroxyapatite
$\mathbb{I}$	Fourth-order identity tensor
k_{HA}	Bulk modulus of hydroxyapatite
k_{H_2O}	Bulk modulus of water
$\mathcal{L}$	Characteristic lengths of geometry or loading of a structure built up by the material defined on the RVE
ℓ	Characteristic length of an RVE
ℓ_{cort}	Characteristic length of an RVE V_{cort} of cortical bone material
ℓ_{ef}	Characteristic length of an RVE $\check{V}_{ef}$ of extrafibrillar space
ℓ_{excel}	Characteristic length of an RVE $\bar{V}_{excel}$ of extracellular bone matrix
ℓ_{exvas}	Characteristic length of an RVE $\tilde{V}_{exvas}$ of extravascular bone material
ℓ_{fib}	Characteristic length of an RVE $\check{V}_{fib}$ of mineralized collagen fibril
ℓ_{wetcol}	Characteristic length of an RVE $\mathring{V}_{col}$ of wet collagen

5 Multiscale Homogenization Theory

$\underline{N}$	Orientation vector aligned with longitudinal axis of hydroxyapatite needle
n_r	Number of material phases within an RVE
$\underline{n}$	Orientation vector perpendicular to $\underline{N}$
RVE	Representative volume element
r	Index denoting a material phase
$\mathbb{P}_r^0$	Fourth-order Hill tensor characterizing the interaction between the phase r and the matrix $\mathbb{C}^0$
$\overset{\circ}{V}_{\text{col}}$	Volume of molecular collagen within an RVE $\overset{\circ}{V}_{\text{wetcol}}$
V_{cort}	Volume of RVE "cortical bone material"
$\check{V}_{\text{ef}}$	Volume of RVE "extrafibrillar space"
$\bar{V}_{\text{ef}}$	Volume of extrafibrillar space within an RVE $\bar{V}_{\text{excel}}$
$\bar{V}_{\text{excel}}$	Volume of RVE "extracellular bone matrix"
$\tilde{V}_{\text{excel}}$	Volume of extracellular bone matrix within an RVE $\tilde{V}_{\text{exvas}}$
$\tilde{V}_{\text{exvas}}$	Volume of RVE "extravascular bone material"
V_{exvas}	Volume of extravascular bone material within an RVE V_{cort}
$\check{V}_{\text{fib}}$	Volume of RVE "mineralized collagen fibril"
$\bar{V}_{\text{fib}}$	Volume of mineralized collagen fibril within an RVE $\bar{V}_{\text{excel}}$
$\bar{V}_{\text{HA}}$	Volume of hydroxyapatite within an RVE $\check{V}_{\text{fib}}$
$\check{V}_{\text{HA}}$	Volume of hydroxyapatite within an RVE $\check{V}_{\text{ef}}$
$\check{V}_{\text{ic}}$	Volume of intercrystalline space within an RVE $\check{V}_{\text{ef}}$
$\overset{\circ}{V}_{\text{im}}$	Volume of intermolecular water within an RVE $\overset{\circ}{V}_{\text{wetcol}}$
$\tilde{V}_{\text{lac}}$	Volume of lacunae within an RVE $\tilde{V}_{\text{exvas}}$
V_{vas}	Volume of Haversian canals within an RVE V_{cort}
$\overset{\circ}{V}_{\text{wetcol}}$	Volume of RVE "wet collagen"
$\check{V}_{\text{wetcol}}$	Volume of wet collagen within an RVE $\check{V}_{\text{fib}}$
$\boldsymbol{\varepsilon}_r$	Second-order "microscopic" strain tensor field within phase r
$\dot{\boldsymbol{\varepsilon}}_r$	Incremental "microscopic" second-order strain tensor field within phase r
$\boldsymbol{\varepsilon}_r^{\text{p}}$	Second-order "microscopic" plastic strain tensor field within phase r
$\dot{\lambda}_r$	Incremental plastic multiplier
ϑ	Latitudinal coordinate of spherical coordinate system
θ	Integration variable, $\theta = 0\ldots\pi$
μ_{HA}	Shear modulus of hydroxyapatite
$\mu_{\text{H}_2\text{O}}$	Shear modulus of water
$\boldsymbol{\sigma}_{\text{col}}$	Second-order stress tensor field within molecular collagen
$\sigma_{\text{col}}^{\text{ult}}$	Uniaxial tensile or compressive strength of molecular collagen
σ_{HA}^{NN}	Normal component of stress tensor $\boldsymbol{\sigma}_{\text{HA}\vartheta\varphi}$ in needle direction
σ_{HA}^{Nn}	Shear component of stress tensor $\boldsymbol{\sigma}_{\text{HA}\vartheta\varphi}$ in planes orthogonal to the needle direction
$\sigma_{\text{HA}}^{\text{ult,s}}$	Uniaxial shear strength of pure HA

$\sigma_{HA}^{ult,t}$	Uniaxial tensile strength of pure HA
σ_r	Second-order stress tensor field within phase r
Σ	Second-order "macroscopic" stress tensor
φ	Longitudinal coordinate of spherical coordinate system
ψ	Longitudinal coordinate of vector $\underline{n}$
$\cdot$	First-order tensor contraction
$:$	Second-order tensor contraction

5.1 Introduction

Biomimetics deals with the application of nature-made "design solutions" to the realm of engineering. In this context, large efforts have aimed at imitating biological materials with interesting mechanical properties. However, biological materials are hierarchically organized and very complex, and frequently, the way they work is not easily comprehensible [1]. Hence, successful biomimetics solutions require a deep understanding of "universal" functioning principles of biological materials. It now appears that multiscale mechanics may hold the key to such an understanding of "building plans" inherent to entire material classes.

For relating the vision of hierarchical organization of materials to effective mechanical properties, we rely on continuum micromechanics, which is a well-established tool for structure–property investigations.

Based on various physical–chemical and mechanical experiments, our focus is the development of multiscale mechanical models, which mathematically and computationally quantify how the basic building blocks of biological materials (such as hydroxyapatite minerals, collagen, and water in all bones found throughout the vertebrate kingdom) govern the materials' mechanical properties at different length scales, from a few nanometers up to the macroscopic level. Thereby, multiscale homogenization theory allows us, at each scale, to identify material representations which are as simple as possible, but as complex as necessary for reliable computational predictions of key material properties, such as poroelasticity, creep, and strength. This can be seen as "reverse" biomimetics engineering: (civil) engineering methods are used to understand biological systems, as described in more detail in Sect. 5.2.

This chapter is mainly devoted to the highly fascinating, hierarchically organized material class "bone" (see Sect. 5.3), and to the "universal" elementary building blocks inherent to this material class (Sect. 5.4). A multiscale micromechanics representation (Sect. 5.5) has opened, for the first time, a profound theoretical understanding of bone mechanics, which is consistent with virtually all major experimental observations, given in more detail in Sect. 5.6. One of our key findings is that bone's mechanical properties are governed by porous polycrystals which the minerals build up as structural complement to the collagen fibrils found in all connective tissues (also in tendon, cartilage, skin). These polycrystals are central not only to the magnitude of elastic anisotropy of bone materials, but also to their tensile-to-compressive strength ratio resulting from universal failure characteristics of differently oriented submicron-sized mineral platelets (Sect. 5.7). Implications of these findings for bone biomaterial design conclude the chapter.

5.2 Fundamentals of Continuum Micromechanics

5.2.1 Representative Volume Element

In continuum micromechanics [2–6], a material is understood as a macrohomogeneous, but microheterogeneous body filling a representative volume element (RVE) with characteristic length ℓ, $\ell \gg d$, d standing for the characteristic length of inhomogeneities within the RVE (see Fig. 5.1), and $\ell \ll \mathcal{L}$, $\mathcal{L}$ standing for the characteristic lengths of geometry or loading of a structure built up by the material defined on the RVE.

In general, the microstructure within one RVE is so complicated that it cannot be described in complete detail. Therefore, quasi-homogeneous subdomains with known physical quantities (such as volume fractions or elastoplastic properties) are reasonably chosen. They are called material phases. The "homogenized" mechanical behavior of the overall material, i.e., the relation between homogeneous deformations acting on the boundary of the RVE and resulting (average) stresses, including the ultimate stresses sustainable by the RVE, can then be estimated from the mechanical behavior of the aforementioned homogeneous phases (representing the inhomogeneities within the RVE), their dosages within the RVE, their

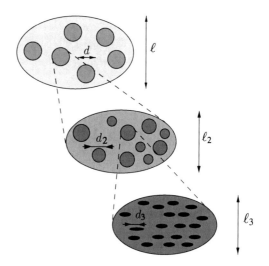

Fig. 5.1 Multistep homogenization: Properties of phases (with characteristic lengths of d and d_2, respectively) inside RVEs with characteristic lengths of ℓ or ℓ_2, respectively, are determined from homogenization over smaller RVEs with characteristic lengths of $\ell_2 \leq d$ and $\ell_3 \leq d_2$, respectively

characteristic shapes, and their interactions. If a single phase exhibits a heterogeneous microstructure itself, its mechanical behavior can be estimated by introduction of an RVE within this phase, with dimensions $\ell_2 \leq d$, comprising again smaller phases with characteristic length $d_2 \ll \ell_2$, and so on, leading to a multistep homogenization scheme (see Fig. 5.1).

5.2.2 Upscaling of Elasto-Brittle and Elastoplastic Material Properties

We consider an RVE consisting of n_r material phases, $r = 1, \ldots, n_r$, exhibiting elastoplastic or elasto-brittle material behavior. In case of ideal associated elastoplasticity, the RVE follows the constitutive laws:

$$\boldsymbol{\sigma}_r = \mathbb{c}_r : (\boldsymbol{\varepsilon}_r - \boldsymbol{\varepsilon}_r^p), \tag{5.1}$$

$$\dot{\boldsymbol{\varepsilon}}_r^p = \dot{\lambda}_r \frac{\partial \mathfrak{f}_r}{\partial \boldsymbol{\sigma}_r}, \quad \dot{\lambda}_r \mathfrak{f}_r(\boldsymbol{\sigma}_r) = 0, \quad \dot{\lambda}_r \geq 0, \quad \mathfrak{f}_r(\boldsymbol{\sigma}_r) \leq 0. \tag{5.2}$$

In (5.2), $\boldsymbol{\sigma}_r$ and $\boldsymbol{\varepsilon}_r$ are the stress and (linearized) strain tensors averaged over phase r with elasticity tensor $\mathbb{c}_r$; $\boldsymbol{\varepsilon}_r^p$ are the average plastic strains in phase r, λ_r is the plastic multiplier of phase r, and $\mathfrak{f}_r(\boldsymbol{\sigma}_r)$ is the yield function describing the (ideally) plastic characteristics of phase r.

In case of brittleness, the RVE follows the constitutive laws:

$$\boldsymbol{\sigma}_r = \mathbb{c}_r : \boldsymbol{\varepsilon}_r \quad (\boldsymbol{\varepsilon}_r^p = \mathbf{0}) \quad \text{if} \quad \mathfrak{f}_r(\boldsymbol{\sigma}_r) < 0, \tag{5.3}$$

$\mathfrak{f}_r(\boldsymbol{\sigma}_r)$ now being a failure function defining stress states $\boldsymbol{\sigma}_r$ related to brittle failure.

The RVE is subjected to Hashin boundary conditions, i.e., to "homogeneous" ("macroscopic") strains $\boldsymbol{E}$ at its boundary, so that the kinematically compatible phase strains $\boldsymbol{\varepsilon}_r$ inside the RVE fulfill the average condition

$$\boldsymbol{E} = \sum_r f_r \boldsymbol{\varepsilon}_r \tag{5.4}$$

with f_r as the volume fraction of phase r. In a similar way, the equilibrated phase stresses $\boldsymbol{\sigma}_r$ fulfill the stress average condition

$$\boldsymbol{\Sigma} = \sum_r f_r \boldsymbol{\sigma}_r \tag{5.5}$$

with $\boldsymbol{\Sigma}$ as the "macroscopic" stresses.

The superposition principle (following from linear elasticity and linearized strain) implies that the phase strains $\boldsymbol{\varepsilon}_r$ are linearly related to both the macroscopic strains E and the free strains ε_r^p (which can be considered as independent loading parameters),

$$\varepsilon_r = \mathbb{A}_r : E + \sum_s \mathbb{D}_{rs} : \varepsilon_s^p \tag{5.6}$$

with $\mathbb{A}_r$ as the fourth-order concentration tensor [7], and $\mathbb{D}_{rs}$ as the fourth-order influence tensors [8]. The latter quantify the phase strains ε_r resulting from plastic strains ε_s^p, while the overall RVE is free from deformation, $E = \mathbf{0}$.

In the absence of plastic strains [$\mathfrak{f}_r < 0$, $\varepsilon_r^p = \mathbf{0}$ in (5.1)–(5.2)], the RVE behaves fully elastically, so that (5.6), (5.5), (5.4), and (5.1) yield a macroscopic elastic law of the form:

$$\Sigma = \mathbb{C}^{\text{hom}} : E \quad \text{with} \quad \mathbb{C}^{\text{hom}} = \sum_r f_r \mathbb{C}_r : \mathbb{A}_r, \tag{5.7}$$

as the homogenized elastic stiffness tensor characterizing the material within the RVE. In case of nonzero "free" plastic strains ε_r^p, (5.7) can be extended to the form:

$$\Sigma = \mathbb{C}^{\text{hom}} : (E - E^p), \tag{5.8}$$

(5.8), together with (5.1), (5.5)–(5.7), gives access to the macroscopic plastic strains E^p, reading as:

$$E^p = -\left[\sum_r f_r \mathbb{C}_r : \mathbb{A}_r\right]^{-1} : \left\{\sum_r f_r \mathbb{C}_r : \left[(\mathbb{A}_r : E + \sum_s \mathbb{D}_{rs} : \varepsilon_s^p) - \varepsilon_r^p\right]\right\} + E. \tag{5.9}$$

Matrix-inclusion problems [9, 10] allow for estimating concentration tensors $\mathbb{A}_r$ and influence tensors $\mathbb{D}_{rs}$ [11], so that the estimate for the homogenized stiffness (5.7) can be written as [5]:

$$\mathbb{C}^{\text{hom}} = \sum_r f_r \mathbb{C}_r : \left[\mathbb{I} + \mathbb{P}_r^0 : (\mathbb{C}_r - \mathbb{C}^0)\right]^{-1} : \left\{\sum_s f_s \left[\mathbb{I} + \mathbb{P}_s^0 : (\mathbb{C}_s - \mathbb{C}^0)\right]^{-1}\right\}^{-1}, \tag{5.10}$$

where $\mathbb{I}$ is the fourth-order unity tensor, and the fourth-order Hill tensor $\mathbb{P}_r^0$ accounts for the characteristic shape of phase r in a matrix with stiffness $\mathbb{C}^0$. The two sums are taken over all phases of the heterogeneous material in the RVE. Choice of this stiffness describes the interactions between the phases: For $\mathbb{C}^0$ coinciding with one of the phase stiffnesses (Mori–Tanaka scheme [12, 13]), a composite material is represented (contiguous matrix with inclusions); for $\mathbb{C}^0 = \mathbb{C}^{hom}$ (self-consistent scheme [2, 14]), a dispersed arrangement of the phases is considered (typical for polycrystals).

5.3 Bone's Hierarchical Organization

Bone materials are characterized by an astonishing variability and diversity. Still, because of "architectural constraints" due to once chosen material constituents and their physical interaction, the fundamental hierarchical organization or basic building plans of bone materials remain largely unchanged during biological evolution. These building plans are expressed by typical morphological features which can be discerned across all bone materials. We here distinguish six levels of hierarchical organization in the line of Katz et al. [15], which have been quite generally accepted in the scientific community:

- At an observation scale of several 10 nm, the so-called elementary components of mineralized tissues can be distinguished. Long cylindrically shaped collagen molecules with a diameter of about 1.2 nm and a length of about 300 nm [16] are self-assembled in staggered organizational schemes with characteristic diameters of 50–500 nm [17–24], and attached to each other at their ends by crosslinks. These cross-linked collagen molecules build up, together with the water-filled intermolecular space, the scale of wet collagen (Fig. 5.2a).

  Needle or plate-shaped mineral crystals, consisting of impure hydroxyapatite (HA; $Ca_{10}[PO_4]_6[OH]_2$) with typical 1–5 nm thickness, and 25–50 nm length [22], are penetrated by the water-filled intercrystalline space (Fig. 5.2c).
- Wet collagen is penetrated by intrafibrillar crystals, forming a collagen–HA network, called fibril (Fig. 5.2b), at an observation scale of about 10 nm.
- The extrafibrillar crystals form aggregates and build up, together with the intercrystalline space, a porous HA polycrystal called extrafibrillar space (Fig. 5.2c), at an observation scale of about 300 nm.
- At an observation scale of several micrometers, the ultrastructure comprises the fibrils (light areas in Fig. 5.2d) and the extrafibrillar space (dark areas in Fig. 5.2d).
- At an observation scale of 100–200 μm, the extravascular bone matrix comprises the ultrastructure and osteocyte-filled cavities called lacunae (Fig. 5.2e).
- At an observation scale of several 100 μm to several millimeters, the microstructure comprises the extravascular bone matrix and the vascular space (Haversian canals) (Fig. 5.2f).

5.4 Elastic and Strength Properties of the Elementary Components of Bone: Hydroxyapatite, Collagen, Water

As regards hydroxyapatite, tests with an ultrasonic interferometer coupled with a solid media pressure apparatus [28, 29] reveal isotropic elastic properties, which are in perfect agreement with recent ab initio calculations of elastic constants of

5 Multiscale Homogenization Theory

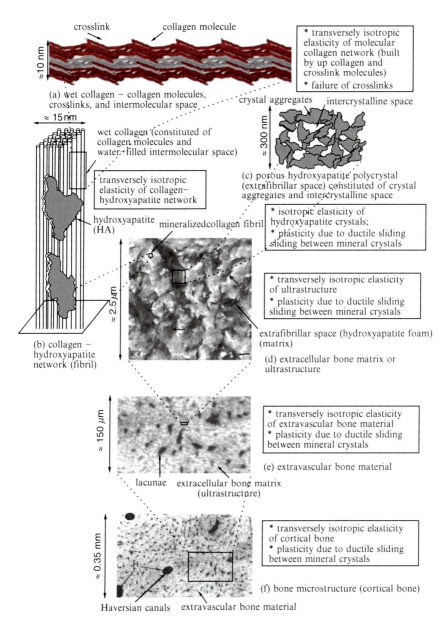

Fig. 5.2 Multiscale view of bone structure, with key physical effects considered in micromechanics representation of Fig. 5.4: (**a**) wet collagen; reproduced from [25], Copyright National Academy of Sciences, USA; (**b**) mineralized collagen fibril; schematic sketch after [26]; (**c**) extrafibrillar porous polycrystal; (**d**) extracellular bone matrix; reproduced with kind permission from Springer Science+Business Media: [24], Fig. 5; (**e**) extravascular bone matrix [zoomed out of image (**f**)]; (**f**) cortical bone; reprinted from [27], with permission from American Institute of Physics, ©1979

Table 5.1 "Universal" (tissue and location-independent) elasticity and strength values of elementary constituents of bone

Material phase	Elasticity (GPa)	Strength (MPa)
	k ... bulk modulus	
	μ ... shear modulus	$\sigma^{ult,t}$ ... uniaxial (tensile) strength
	c_{ijkl} ... tensor components, base frame according to Fig. 5.4	$\sigma^{ult,s}$ ... shear strength
Hydroxyapatite (experimental source)	$k_{HA} = 82.6$ [28] $\mu_{HA} = 44.9$ [28]	$\sigma_{HA}^{ult,t} = 52.2$ [32,33] $\sigma_{HA}^{ult,s} = 80.3$ [32,33]
Water containing noncollagenous organics or osteocytes	$k_{H_2O} = 2.3$ [35] $\mu_{H_2O} = 0$	
Collagen (experimental source)	$c_{col,3333} = 17.9$ $c_{col,1111} = 11.7$ $c_{col,1133} = 7.1$ $c_{col,1122} = 5.1$ $c_{col,1313} = 3.3$ [17]	$\sigma_{col}^{ult} = 144.7$ [36,37]

ceramic crystals [30,31] (Table 5.1); and quasi-static mechanical tests in uniaxial tension and compression on fairly dense samples of artificial hydroxyapatite biomaterials [32,33] give access to the uniaxial tensile and shear strength of pure hydroxyapatite (Table 5.1). The latter is probably governed by detachment of one nonspherical nano- or microcrystal of hydroxyapatite from its neighbors [34]. This is mathematically expressed through [11]:

$$\psi = 0, \ldots, 2\pi :$$

$$f_{HA\varphi\vartheta}(\sigma_{HA\varphi\vartheta}) = \frac{\sigma_{HA}^{ult,t}}{\sigma_{HA}^{ult,s}} \max_{\psi} |\sigma_{HA}^{Nn}| + \sigma_{HA}^{NN} - \sigma_{HA}^{ult,t} = 0 \quad (5.11)$$

with spherical coordinates φ and ϑ defining the crystal needle orientation vector $\underline{N} = \underline{e}_r$ in the reference frame ($\underline{e}_1, \underline{e}_2, \underline{e}_3$), and with ψ defining the orientation of vector $\underline{n}$ related to shear stresses (see Fig. 5.3), with $\sigma_{HA}^{ult,t}$ as the uniaxial tensile strength and $\sigma_{HA}^{ult,s}$ as the shear strength of pure hydroxyapatite; and with $\sigma_{HA}^{Nn} = \underline{N} \cdot \boldsymbol{\sigma}_{HA\varphi\vartheta} \cdot \underline{n}$ and $\sigma_{HA}^{NN} = \underline{N} \cdot \boldsymbol{\sigma}_{HA\varphi\vartheta} \cdot \underline{N}$ as the normal and shear stress components related to a surface with normal $\underline{N}(\varphi, \vartheta)$.

As regards (molecular) collagen, its (transversely isotropic) elastic properties are well accessible through those of dry rat tail tendon, a tissue consisting almost exclusively of collagen (i.e., the intermolecular porous space, which is increasing with the water content of "wet collagen," is minimal for "dry collagen" [16,37]). Corresponding components of the elasticity tensor of (molecular) collagen (see Table 5.1), derived from Brillouin light scattering tests [17], are in perfect agreement with Young's modulus of a collagen I monomer, when measured through an optical

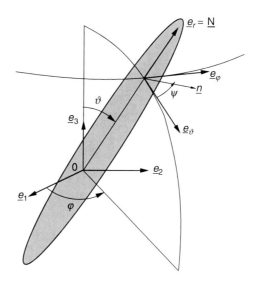

Fig. 5.3 Cylindrical (needle-like) HA inclusion oriented along vector $\underline{N}$ and inclined by angles ϑ and φ with respect to the reference frame $(\underline{e}_1, \underline{e}_2, \underline{e}_3)$; local base frame $(\underline{e}_r, \underline{e}_\vartheta, \underline{e}_\varphi)$ is attached to the needle

tweezer-interferometer system [38]. Strength tests, also on rat tail tendon [36], reveal the crosslink-failure-induced quasi-brittle strength of collagen [39], when related to the molecular collagen portion of the tissue [11] (see Table 5.1). The latter is mathematically expressed through [11]:

$$f_{col}(\boldsymbol{\sigma}_{col}) = |\underline{e}_3 \cdot \boldsymbol{\sigma}_{col} \cdot \underline{e}_3| - \sigma_{col}^{ult} \leq 0, \quad (5.12)$$

where the direction three coincides with the principal orientation direction of collagen and where σ_{col}^{ult} is the uniaxial (tensile and compressive) strength of molecular collagen (see Fig. 5.4).

We assign the ultrasonics-derived bulk modulus of water [35] (and no strength value) to phases comprising water with mechanically insignificant noncollagenous organic matter (Table 5.1).

5.5 Multiscale Micromechanical Representation of Bone

Across the hierarchical organization of cortical bone material, the following "universal" microstructural patterns are considered in the framework of a multistep homogenization scheme (Fig. 5.4, see [11,40,41] for details): The first homogenization step refers to an observation scale of several nanometers, where cross-linked collagen molecules form a (linear elastic-perfectly brittle) contiguous matrix, which is "perforated" by intermolecular, water-filled spaces. A Mori–Tanaka scheme relates the stiffnesses and volume fractions of molecular collagen and intermolecular

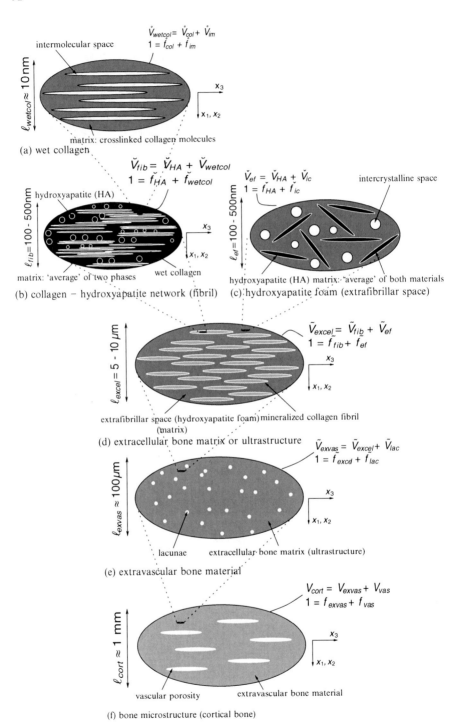

Fig. 5.4 Micromechanical representation of bone material by means of a six-step homogenization procedure. Reproduced from [11]

space to the homogenized elasticity tensor of wet collagen [specification of (5.10) for RVE of Fig. 5.4a]. The second homogenization step refers to an observation scale of several hundreds of nanometers, where wet collagen and mineral crystal agglomerations penetrate each other, building up the mineralized collagen fibril. A self-consistent scheme relates the stiffnesses and volume fractions of wet collagen and (intrafibrillar) hydroxyapatite crystal agglomerations to the homogenized elasticity tensor of the fibril [specification of (5.10) for RVE of Fig. 5.4b]. The third homogenization step also refers to an observation scale of several hundreds of nanometers, where (extrafibrillar) hydroxyapatite crystal agglomerations and intercrystalline space build up a porous polycrystal (extrafibrillar space). A self-consistent scheme relates the stiffnesses and volume fractions of hydroxyapatite crystals and intercrystalline space to the homogenized elasticity tensor of the porous hydroxyapatite crystal [specification of (5.10) for RVE of Fig. 5.4c]. The fourth homogenization step refers to an observation scale of some micrometers, where mineralized fibrils are embedded as inclusions into the extrafibrillar mineral foam, forming together the extracellular bone matrix (ultrastructure). A Mori–Tanaka scheme relates the stiffnesses and volume fractions of the extrafibrillar space and the fibril to the homogenized elasticity tensor of the extracellular bone matrix [specification of (5.10) for RVE of Fig. 5.4d]. The fifth homogenization step refers to an observation scale of about $100\,\mu$m, where lacunar pores are embedded in an ultrastructural matrix building up the extravascular bone material. A Mori–Tanaka scheme relates the stiffnesses and volume fractions of the ultrastructure and the lacunae to the homogenized elasticity tensor of the extravascular bone material [specification of (5.10) for RVE of Fig. 5.4e]. The sixth homogenization step refers to an observation scale of about 1 mm, where the extravascular matrix is perforated by Haversian canals, building up the microstructure. A Mori–Tanaka scheme relates the stiffnesses and volume fractions of the extravascular bone material and the Haversian canals to the homogenized elasticity tensor of the microstructure [specification of (5.10) for RVE of Fig. 5.4f].

In addition, repeated specification of (5.6) for all RVEs in Fig. 5.4 allows for upscaling strains from the level of the elementary components, all the way up to the macroscopic bone material of Fig. 5.4f. Considering, in addition, the elastoplastic behavior of sliding hydroxyapatite minerals [11, 42], according to (5.11), (5.1), and (5.2), the elastoplastic material behavior at all scales is defined, inclusive of quasi-brittle failure once the collagen failure surface, (5.12), is reached.

5.6 Experimental Validation of Multiscale Micromechanics Theory for Bone

The micromechanical model of Sect. 5.5 predicts, for each set of tissue-specific volume fractions, the corresponding tissue-specific elasticity and strength properties at all observation scales of Fig. 5.4. Thus, a strict experimental validation of this model is realized as follows: (1) different sets of *volume fractions* [variables f_r in (5.1)–(5.10)] are determined from composition experiments on different bone samples with different ages, from different species and different anatomical locations; (2) these volume fractions are used as model input, and (3) corresponding model-predicted *stiffness and strength values* [model output, (5.10)] are compared to results from stiffness and strength experiments on the same or very similar bone samples.

The *volume fractions* within each of the considered RVEs (Fig. 5.4) are accessible through the following experiments: At the microstructural and the extravascular observation scales (Fig. 5.4e,f), polarized light microscopy (Fig. 5.2f) visualizes, in cortical bone, Haversian canals and lacunae [27, 43, 44], from which the vascular and lacunar porosities can be derived, see [41] for details. For determination of the volume fractions in all other RVEs, the pioneering volume measurements and weighing experiments on wet, dehydrated, and demineralized cortical bone specimens, performed by Lees [16, 45, 46], are of central importance; they give access to the mineral, (molecular) collagen, and water content at the extracellular scale (Fig. 5.2d and RVE Fig. 5.4d). The collagen volume fraction in a piece of extracellular bone material (Fig. 5.2d and Fig. 5.4d), together with the spatial organization of collagen molecules within the extracellular matrix, characterized by a staggered longitudinal arrangement quantified by the Hodge–Petruska scheme [47] and by a quasi-hexagonal transverse arrangement as elucidated by the pioneering neutron diffraction experiments of Lees and coworkers [16, 37], gives access to the volume fraction of collagen fibrils in an extracellular RVE (Fig. 5.4d, see [41] for details). Concerning the RVEs below the extracellular scale (Fig. 5.4b,c), their mineral volume fraction can be determined from Hellmich and Ulm's finding [48] that the hydroxyapatite concentration in the extra-collagenous space is the same inside and outside the fibrils, see [41] for details. The class of porous hydroxyapatite biomaterials [32, 33, 49–52] directly mimics the porous polycrystal of Fig. 5.4c. For such materials, referred to in Fig. 5.5a–d, the mineral volume fraction is directly related to porosity or apparent mass density, accessible from mass and volume measurements on biomaterial samples. Finally, the ultrastructural collagen volume and that of wet collagen in fibrils (Fig. 5.4b) give access to the intermolecular porosity within an RVE of wet collagen (see Fig. 5.4a and [41] for details).

Stiffness and strength values are gained from ultrasonics experiments on cortical [43–46, 53] and on trabecular bone samples [54], and from quasi-static uniaxial, compressive or tensile, mechanical tests [42, 55–71] (see [11, 34, 41] for details).

Throughout all different observation scales, and across a great variety of bone tissues (from different anatomical locations of different mammals including humans, cows, elephants, deers, rabbits, and horses), micromechanics model predictions for stiffness and strength agree very well with corresponding experimental data (see Fig. 5.5). This unparalleled quantification of bone's mechanical behavior allows us to draw basic conclusions on how bone works mechanically – described in the next chapter.

5 Multiscale Homogenization Theory

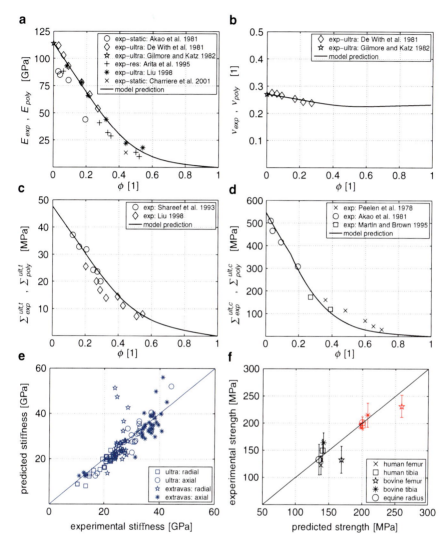

Fig. 5.5 Comparison between micromechanical model predictions and corresponding experiments: (**a, b**) elasticity in terms of Young's modulus E and Poisson's ratio ν of hydroxyapatite biomaterials (see Fig. 5.4c, [34]); (**c, d**) uniaxial tensile and compressive strength of hydroxyapatite biomaterials (see Fig. 5.4c, [34]); (**e**) radial and axial normal stiffness (C_{1111} and C_{3333}) of cortical and trabecular bone at ultrastructural (ultra) and extravascular (exvas) scales (see Fig. 5.4d,e, [41]); (**f**) uniaxial strength of cortical bone at the macroscopic scale (see Fig. 5.4f, [11]), with mean and standard deviation being depicted for experimental tensile strength (*dark color*) and experimental compressive strength (*light color*)

5.7 How Bone Works: Mechanical Design Characteristics of Bone Revealed Through Multiscale Micromechanics

Bone materials are built up in a remarkably anisotropic fashion, with respect to both stiffness and strength.

As regards elasticity, the experimentally verified micromechanics theory of Sects. 5.5 and 5.6 reveals the following sources of anisotropy:

1. The intrinsic anisotropy of molecular collagen, quantified through Brillouin light scattering tests of Cusack and Miller [17] and given at the bottom of Table 5.1;
2. The axially oriented arrangement of molecular collagen in the mineralized collagen fibril (Fig. 5.4b); of the mineralized collagen fibrils in the extracellular bone matrix (Fig. 5.4d); and of the vascular porosity (Haversian canals) in cortical bone (Fig. 5.4f).

In contrast, bone mineral is poorly organized [20, 24, 72–75], so that bone materials without collagen and vascular pores, such as hyperpycnotic tissues [16,76] or artificial hydroxyapatite biomaterials [29,32,51,52,77,78], represented basically through the extrafibrillar RVE of Fig. 5.4c, are virtually isotropic. At the same time, the high connectivity of the individual crystals, forming a porous polycrystal in the extrafibrillar space of bone materials, is important for providing sufficient transverse stiffness.

As regards strength, the micromechanics model of Sects. 5.5 and 5.6 shows an additional feature beyond anisotropy, namely the sensitivity of bone to tensile versus compressive loading type (not existing in the elastic regime): The extrafibrillar space (see Fig. 5.2c and RVE in Fig. 5.4c) is ten times weaker under tensile loading than under compressive loading, a characteristic which it shares with a lot of other quasi-brittle materials, such as sandstone or concrete [79]. The reason for that lies in the localization or concentration of homogeneous tensile or compressive strains from the boundary of the extrafibrillar RVE into its individual needle phases, leading eventually to failure of the most unfavorably stressed needle according to criterion (5.11) (see Fig. 5.6). In overall uniaxial tension, this local failure is dominated by tensile normal stresses in needle phases more or less aligned with the loading direction (for more or less porous materials; see Fig. 5.6a,b). In overall uniaxial compression, however, the aforementioned local failure is still dominated by tensile normal stresses in one needle phase – this phase now being aligned more or less perpendicular to the loading direction (for more or less porous materials; see Fig. 5.6c,d). Accordingly, it takes much more overall compressive stresses to produce as much normal tensile stresses in needles perpendicular to the loading direction than tensile stresses in needles aligned with the loading direction, stemming from overall tensile loads. This explains the ratio of 1:10 between tensile and compressive strength in hydroxyapatite biomaterials.

In real bones, however, two differences with respect to this situation are significant:

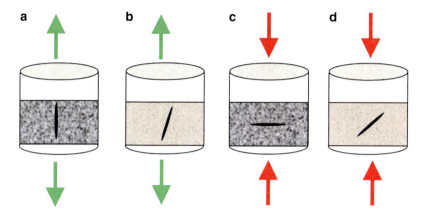

Fig. 5.6 Orientation of crystal needle initiating overall failure of hydroxyapatite polycrystals, under tensile (**a, b**) and compressive (**c, d**) uniaxial macroscopic loading: (**a, c**) local tensile failure in highly porous polycrystals; (**b, d**) local combined tensile-shear failure in polycrystals with low porosity

1. The hydroxyapatite crystals are nanometer-sized instead of micrometer-sized, so that they are intimately bond to the structured water layers between them [80–83]. This evokes ideal plastic sliding rather than brittle failure, once the elastic regime of the crystals is left.
2. Molecular collagen (shown to be quasi-brittle, both experimentally [36] and computationally [84]) reinforces the multiporous hierarchical nanocomposite "bone" both in axial tension and in axial compression, up to the same extent. Accordingly, bone materials show a tensile-to-compressive strength ratio of 2:3, instead of 1:10.

Bone replacement materials should mimic these "microstructural mechanics", features as closely as possible if an imitation of the natural form of bone is desired [105]. Hence, the design characteristics of natural bone imply the following design rules for creating bone replacement materials close to their natural "prototypes":

- Use of an anisotropic organic template
- Growth of nanometer-sized mineral crystals
- Realization of a connected network of crystals

In this context, the following recent developments in the bone biomaterials community seem noteworthy: As regards anisotropic organic templates, recent bone biomaterials rely on materials such as starch [85], chitosan [86], or collagen itself [87–92]. While micron-sized hydroxyapatite crystals defined the state of the art in biomaterial production in the 1980s and 1990s [32, 33, 50, 77], modern processing techniques, such as sol–gel combustion [93] or precipitation from an aqueous solution [87,88,92,94–97], allow for production of nanometer-sized hydroxyapatite crystals. Finally, also interconnected networks of hydroxyapatite crystals could be recently produced, by means of a layer-by-layer deposition method [92].

5.8 Some Conclusions from a Biological Viewpoint

To the knowledge of the authors, the aforementioned design rules concern all different kinds of bones, across the entire subphylum *vertebrata*, in the sense that they satisfactorily explain, in a unified fashion, the large variety of mechanical properties depicted in Fig. 5.5e,f. In this sense, they are "universal" for the vertebrates, but not beyond this subphylum. The aforementioned design principles may result from "architectural constraints" [98, 99] merely due to once chosen material constituents and their physical interactions that imply the fundamental hierarchical organization patterns or basic building plans of bone materials (see Fig. 5.2), which remain largely unchanged during biological evolution. In this sense, it seems highly probable that the same principles hold for extinct species like dinosaurs, although this cannot be tested in full completeness, as fossils are chemically altered postmortem [100].

Mineralized skeletal structures built up of bone typically span three orders of magnitude (Fig. 5.7), whereby the mammals currently define the upper size limit. The concept of an inner skeleton ("endoskeleton") is also followed out of the phylum *chordata*, namely with sponges (using either silica or calcium carbonates as mineral, Fig. 5.7) and with phylum *echinodermata*, based on mineralized tissues containing calcium carbonates (Fig. 5.7). The same holds for certain members of phylum *mollusca*, namely cephalopods (Fig. 5.7). Alternatively, load-carrying organs may be realized as outer skeletons (exoskeletons), at tiny formats containing silica (e.g., in diatoms), and at larger formats containing calcium carbonates and chitin as organic component (with phylum *arthropoda* and with all classes of phylum *mollusca* except for certain cephalopods, see Fig. 5.7).

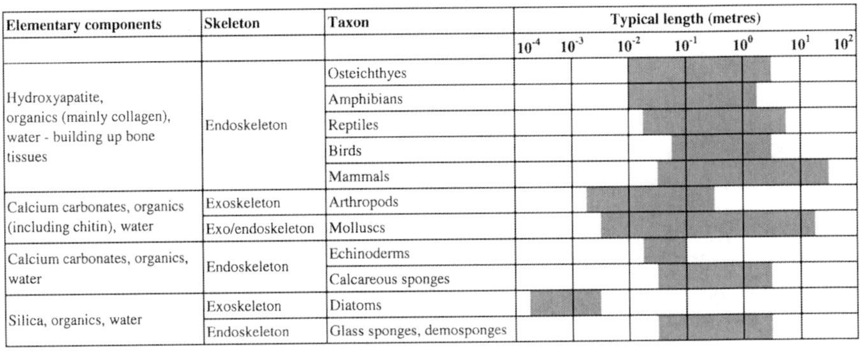

Elementary components	Skeleton	Taxon	Typical length (metres) 10^{-4} – 10^{2}
Hydroxyapatite, organics (mainly collagen), water - building up bone tissues	Endoskeleton	Osteichthyes	
		Amphibians	
		Reptiles	
		Birds	
		Mammals	
Calcium carbonates, organics (including chitin), water	Exoskeleton	Arthropods	
	Exo/endoskeleton	Molluscs	
Calcium carbonates, organics, water	Endoskeleton	Echinoderms	
		Calcareous sponges	
Silica, organics, water	Exoskeleton	Diatoms	
	Endoskeleton	Glass sponges, demosponges	

Fig. 5.7 Overview over elementary components of mineralized tissues and size of inner and outer skeletons built thereof, across the animal kingdom (see also [101–104])

References

1. Ch. Hellmich, D. Katti, Mechanics of biological and bioinspired materials and structures. J. Eng. Mech ASCE **35**(5), 365–366 (2009)
2. R. Hill, Elastic properties of reinforced solids: some theoretical principles. J. Mech. Phys. Solids **11**, 357–362 (1963)
3. P. Suquet, Effective behavior of nonlinear composites, in *Continuum Micromechanics*, ed. by P. Suquet (Springer, Wien, New York, 1997), pp. 197 – 264
4. A. Zaoui, Structural morphology and constitutive behavior of microheterogeneous materials, in *Continuum Micromechanics*, ed. by P. Suquet (Springer, Wien, New York, 1997), pp. 291–347
5. A. Zaoui, Continuum micromechanics: survey. J. Eng. Mech ASCE **128**(8), 808–816 (2002)
6. L. Dormieux, D. Kondo, F.-J. Ulm, *Microporomechanics* (Wiley, 2006)
7. R. Hill, Continuum micro-mechanics of elastoplastic polycrystals. J. Mech. Phys. Solids **13**, 89–101 (1965)
8. G.J. Dvorak, Transformation field analysis of inelastic composite materials. Proc. R. Soc. Lond. A **437**, 311–327 (1992)
9. J.D. Eshelby, The determination of the elastic field of an ellipsoidal inclusion, and related problems. Proc. R. Soc. Lond. A **241**, 376–396 (1957)
10. N. Laws, The determination of stress and strain concentrations at an ellipsoidal inclusion in an anisotropic material. J. Elasticity **7**(1), 91–97 (1977)
11. A. Fritsch, Ch. Hellmich, L. Dormieux, Ductile sliding between mineral crystals followed by rupture of collagen crosslinks: experimentally supported micromechanical explanation of bone strength. J. Theor. Biol. **260**, 230–252 (2009)
12. T. Mori, K. Tanaka, Average stress in matrix and average elastic energy of materials with misfitting inclusions. Acta Metallurgica **21**(5), 571–574 (1973)
13. K. Wakashima, H. Tsukamoto, Mean-field micromechanics model and its application to the analysis of thermomechanical behaviour of composite materials. Mater. Sci. Eng. A **146**(1–2), 291–316 (1991)
14. A.V. Hershey, The elasticity of an isotropic aggregate of anisotropic cubic crystals. J. Appl. Mech. ASME **21**, 236–240 (1954)
15. J.L. Katz, H.S. Yoon, S. Lipson, R Maharidge, A. Meunier, P. Christel, The effects of remodelling on the elastic properties of bone. Calcif. Tissue Int **36**, S31–S36 (1984)
16. S. Lees, Considerations regarding the structure of the mammalian mineralized osteoid from viewpoint of the generalized packing model. Connect. Tissue Res. **16**, 281–303 (1987)
17. S. Cusack, A. Miller, Determination of the elastic constants of collagen by Brillouin light scattering. J. Mol. Biol. **135**, 39–51 (1979)
18. A. Miller, Collagen: the organic matrix of bone. Philos Trans. R. Soc. Lond. B **304**, 455–477 (1984)
19. S. Lees, N.-J. Tao, M. Lindsay, Studies of compact hard tissues and collagen by means of Brillouin light scattering. Connect. Tissue Res. **24**, 187–205 (1990)
20. S. Lees, K.S. Prostak, V.K. Ingle, K. Kjoller, The loci of mineral in turkey leg tendon as seen by atomic force microscope and electron microscopy. *Calcif. Tissue Int.* **55**, 180–189 (1994)
21. S. Weiner, T. Arad, I. Sabanay, W. Traub, Rotated plywood structure of primary lamellar bone in the rat: orientation of the collagen fibril arrays. Bone **20**, 509–514 (1997)
22. S. Weiner, H.D. Wagner, The material bone: structure – mechanical function relations. Annu. Rev. Mater. Sci. **28**, 271–298 (1998)
23. J.-Y. Rho, L. Kuhn-Spearing, P. Zioupos, Mechanical properties and the hierarchical structure of bone. Med. Eng. Phys. **20**, 92–102 (1998)
24. K.S. Prostak, S. Lees, Visualization of crystal-matrix structure. In situ demineralization of mineralized turkey leg tendon and bone. Calcified Tissue Int. **59**, 474–479 (1996)
25. J.P.R.O. Orgel, T.C. Irving, A. Miller, T.J. Wess, Microfibrillar structure of type I collagen in situ. Proc. Natl. Acad. Sci. USA **103**(24), 9001–9005 (2006)

26. W.J. Landis, M.J. Song, A. Leith, L. McEwen, B.F. McEwen, Mineral and organic matrix interaction in normally calcifying tendon visualized in three dimensions by high-voltage electron microscopic tomography and graphic image reconstruction. J. Struct. Biol. **110**, 39–54 (1993)
27. S. Lees, P. Cleary, J.D. Heeley, E.L. Gariepy, Distribution of sonic plesio-velocity in a compact bone sample. J. Acoust. Soc. Am. **66**(3), 641–646 (1979)
28. J.L. Katz, K. Ukraincik, On the anisotropic elastic properties of hydroxyapatite. J. Biomech. **4**, 221–227 (1971)
29. R.S. Gilmore, J.L. Katz, Elastic properties of apatites. J. Mater. Sci. **17**, 1131–1141 (1982)
30. H. Yao, L. Ouyang, W.-Y. Ching, Ab initio calculation of elastic constants of ceramic crystals. J. Am. Ceramic Soc. **90**(10), 3194–3204 (2007)
31. W.Y. Ching, P. Rulis, A. Misra, Ab initio elastic properties and tensile strength of crystalline hydroxyapatite. Acta Biomater. **5**, 3067–3075 (2009)
32. M. Akao, H. Aoki, K. Kato, Mechanical properties of sintered hydroxyapatite for prosthetic applications. J. Mater. Sci. **16**, 809–812 (1981)
33. M.Y. Shareef, P.F. Messer, R. van Noort, Fabrication, characterization and fracture study of a machinable hydroxyapatite ceramic. Biomaterials **14**(1), 69–75 (1993)
34. A. Fritsch, L. Dormieux, Ch. Hellmich, J. Sanahuja, Mechanical behaviour of hydroxyapatite biomaterials: an experimentally validated micromechanical model for elasticity and strength. J. Biomed. Mater. Res. A **88A**, 149–161 (2009)
35. N. Bilaniuk, G.S.K. Wong, Speed of sound in pure water as a function of temperature. J. Acoust. Soc. Am. **93**(3), 1609–1612 (1993)
36. E. Gentleman, A.N. Lay, D.A. Dickerson, E.A. Nauman, G.A. Livesay, K.C. Dee. Mechanical characterization of collagen fibers and scaffolds for tissue engineering. Biomaterials **24**, 3805–3813 (2003)
37. S. Lees, L.C. Bonar, H.A. Mook, A study of dense mineralized tissue by neutron diffraction. Int. J. Biol. Macromol. **6**, 321–326 (1984)
38. Y.-L. Sun, Z.-P. Luo, A. Fertala, K.-N. An, Direct quantification of the flexibility of type I collagen monomer. Biochem. Biophys. Res. Commun. **295**, 382–386 (2002)
39. M.J. Buehler, Nanomechanics of collagen fibrils under varying cross-link densities: atomistic and continuum studies. J. Mech. Behav. Biomed. Mater. **1**, 59–67 (2008)
40. Ch. Hellmich, J.-F. Barthélémy, L. Dormieux, Mineral-collagen interactions in elasticity of bone ultrastructure – a continuum micromechanics approach. Eur. J. Mech. A Solids **23**, 783–810 (2004)
41. A. Fritsch, Ch. Hellmich, 'Universal' microstructural patterns in cortical and trabecular, extra-cellular and extravascular bone materials: micromechanics-based prediction of anisotropic elasticity. J. Theor. Biol. **244**(4), 597–620 (2007)
42. A.H. Burstein, J.J.M. Zika, K.G. Heiple, L. Klein, Contribution of collagen and mineral to the elastic-plastic properties of bone. J. Bone Joint Surg. **57A**, 956–961 (1975)
43. S. Lees, D. Hanson, E.A. Page, H.A. Mook, Comparison of dosage-dependent effects of beta-aminopropionitrile, sodium fluoride, and hydrocortisone on selected physical properties of cortical bone. J. Bone Miner. Res. **9**(9), 1377–1389 (1994)
44. R.N. McCarthy, L.B. Jeffcott, R.N. McCartney, Ultrasound speed in equine cortical bone: effects of orientation, density, porosity and temperature. J. Biomech. **23**(11), 1139–1143 (1990)
45. S. Lees, J.D. Heeley, P.F. Cleary, A study of some properties of a sample of bovine cortical bone using ultrasound. Calcif. Tissue Int. **29**, 107–117 (1979)
46. S. Lees, J.M. Ahern, M. Leonard, Parameters influencing the sonic velocity in compact calcified tissues of various species. J. Acoust. Soc. Am. **74**(1), 28–33 (1983)
47. A.J. Hodge, J.A. Petruska, Recent studies with the electron microscope on ordered aggregates of the tropocollagen molecule, in *Aspects of Protein Structure – Proceedings of a Symposium Held in Madras 14–18 January 1963 and Organized by the University of Madras, India*, ed. by G.N. Ramachandran (Academic, London, 1963), pp. 289–300

48. Ch. Hellmich, F.-J. Ulm, Average hydroxyapatite concentration is uniform in extracollagenous ultrastructure of mineralized tissue. Biomech. Model. Mechanobiol. **2**, 21–36 (2003)
49. J.G.J. Peelen, B.V. Rejda, K. de Groot, Preparation and properties of sintered hydroxylapatite. Ceramurgia Int. **4**(2), 71–74 (1978)
50. R.I. Martin, P.W. Brown, Mechanical properties of hydroxyapatite formed at physiological temperature. J. Mater. Sci. Mater. Med. **6**, 138–143 (1995)
51. D.-M. Liu, Preparation and characterisation of porous hydroxyapatite bioceramic via a slip-casting route. Ceramics Int. **24**, 441–446 (1998)
52. E. Charrière, S. Terrazzoni, C. Pittet, Ph. Mordasini, M. Dutoit, J. Lemaître, Ph. Zysset, Mechanical characterization of brushite and hydroxyapatite cements. Biomaterials **22**, 2937–2945 (2001)
53. R.B. Ashman, S.C. Cowin, W.C. van Buskirk, J.C. Rice, A continuous wave technique for the measurement of the elastic properties of cortical bone. J. Biomech. **17**(5), 349–361 (1984)
54. R.B. Ashman, J.Y. Rho, Elastic modulus of trabecular bone material. J. Biomech. **21**(3), 177–181 (1988)
55. J.D. Currey, Differences in the tensile strength of bone of different histological types. J. Anat. **93**, 87–95 (1959)
56. E.D. Sedlin, C. Hirsch, Factors affecting the determination of the physical properties of femoral cortical bone. Acta Orthop. Scand. **37**, 29–48 (1966)
57. A.H. Burstein, J.D. Currey, V.H. Frankel, D.T. Reilly, The ultimate properties of bone tissue: the effects of yielding. J. Biomech. **5**, 35–44 (1972)
58. D.T. Reilly, A.H. Burstein, The elastic modulus for bone. J. Biomech. **7**, 271–275 (1974)
59. D.T. Reilly, A.H. Burstein, The elastic and ultimate properties of compact bone tissue. J. Biomech. **8**, 393–405 (1975)
60. J.D. Currey, The effects of strain rate, reconstruction and mineral content on some mechanical properties of bovine bone. J. Biomech. **8**, 81–86 (1975)
61. A.H. Burstein, D.T. Reilly, M. Martens, Aging of bone tissue: mechanical properties. J. Bone Joint Surg. **58A**, 82–86 (1976)
62. R.P. Dickenson, W.C. Hutton, J.R. Stott, The mechanical properties of bone in osteoporosis. J. Bone Joint Surg. **63-B**(2), 233–238 (1981)
63. H. Cezayirlioglu, E. Bahniuk, D.T. Davy, K.G. Heiple, Anisotropic yield behavior of bone under combined axial force and tension. J. Biomech. **18**(1), 61–69 (1985)
64. R.B. Martin, J. Ishida, The relative effects of collagen fiber orientation, porosity, density, and mineralization on bone strength. J. Biomech. **22**, 419–426 (1989)
65. J.D. Currey, Physical characteristics affecting the tensile failure properties of compact bone. J. Biomech. **23**, 837–844 (1990)
66. R.W. McCalden, J.A. McGeough, M.B. Barker, C.M. Court-Brown, Age-related changes in the tensile properties of cortical bone. The relative importance of changes in porosity, mineralization, and microstructure. J. Bone Joint Surg. **75-A**(8), 1193–1205 (1993)
67. C.M. Riggs, L.C. Vaughan, G.P. Evans, L.E. Lanyon, A. Boyde, Mechanical implications of collagen fibre orientation in cortical bone of the equine radius. Anat. Embryol. **187**, 239–248 (1993)
68. S.C. Lee, B.S. Coan, M.L. Bouxsein, Tibial ultrasound velocity measured in situ predicts the material properties of tibial cortical bone. Bone **21**(1), 119–125 (1997)
69. S.P. Kotha, N. Guzelsu, Modeling the tensile mechanical behavior of bone along the longitudinal direction. J. Theor. Biol. **219**, 269–279 (2002)
70. J.D. Currey, Tensile yield in compact bone is determined by strain, post-yield behaviour by mineral content. J. Biomech. **37**, 549–556 (2004)
71. Ch. Hellmich, H.W. Müllner, Ch. Kohlhauser, Mechanical (triaxial) tests on biological materials and biomaterials. Technical Report DNRT3-1.2-3, Network of Excellence 'Knowledge-based Multicomponent Materials for Durable and Safe Performance – KMM-NoE', sponsored by the European Commission, October 2006
72. F. Peters, K. Schwarz, M. Epple, The structure of bone studied with synchrotron X-ray diffraction, X-ray absorption spectroscopy and thermal analysis. Thermochim. Acta **361**, 131–138 (2000)

73. M. Epple, Solid-state chemical methods to investigate the nature of calcified deposits. Zeitschrift für Kardiologie **90**(Suppl. 3), III/64–III/67 (2001)
74. V. Benezra Rosen, L.W. Hobbs, M. Spector, The ultrastructure of anorganic bovine bone and selected synthetic hydroxyapatites used as bone graft substitute material. Biomaterials **23**, 921–928 (2002)
75. Ch. Hellmich, F.-J. Ulm, Are mineralized tissues open crystal foams reinforced by crosslinked collagen? Some energy arguments. J. Biomech. **35**, 1199–1212 (2002)
76. S. Lees, D. Hanson, E.A. Page, Some acoustical properties of the otic bones of a fin whale. J. Acoust. Soc. Am. **99**(4), 2421–2427 (1995)
77. G. De With, H.J.A. van Dijk, N. Hattu, K. Prijs, Preparation, microstructure and mechanical properties of dense polycrystalline hydroxy apatite. J. Mater. Sci. **16**, 1592–1598 (1981)
78. I.H. Arita, D.S. Wilkinson, M.A. Mondragón, V.M. Castaño, Chemistry and sintering behaviour of thin hydroxyapatite ceramics with controlled porosity. Biomaterials **16**, 403–408 (1995)
79. H. Kupfer, H.K. Hilsdorf, H. Rusch, Behavior of concrete under biaxial stresses. ACI J. **66**, 656–666 (1969)
80. D. Zahn, O. Hochrein, Computational study of interfaces between hydroxyapatite and water. Phys. Chem. Chem. Phys. **5**, 4004–4007 (2003)
81. D. Zahn, O. Hochrein, A. Kawska, J. Brickmann, R. Kniep, Towards an atomistic understanding of apatite-collagen biomaterials: linking molecular simulation studies of complex-, crystal- and composite-formation to experimental findings. J. Mater. Sci. **42**, 8966–8973 (2007)
82. R. Bhowmik, K.S. Katti, D.R. Katti, Mechanics of molecular collagen is influenced by hydroxyapatite in natural bone. J. Mater. Sci. **42**, 8795–8803 (2007)
83. R. Bhowmik, K.S. Katti, D.R. Katti, Mechanisms of load-deformation behavior of molecular collagen in hydroxyapatite-tropocollagen molecular system: steered molecular dynamics study. J. Eng. Mech. **135**(5), 413–421 (2009)
84. M.J. Buehler, Nature designs tough collagen: explaining the nanostructure of collagen fibrils. Proc. Natl. Acad. Sci. USA **103**(33), 12285–12290 (2006)
85. J.F. Mano, C.M. Vaz, S.C. Mendes, R.L. Reis, A.M. Cunha, Dynamic mechanical properties of hydroxyapatite-reinforced and porous starch-based degradable biomaterials. J. Mater. Sci. Mater. Med. **10**, 857–862 (1999)
86. D. Verma, K. Katti, D. Katti, Effect of biopolymers on structure of hydroxyapatite and interfacial interactions in biomimetically synthesized hydroxyapatite/biopolymer nanocomposites. Ann. Biomed. Eng. **36**(6), 1024–1032 (2008)
87. C. Du, F.Z. Cui, X.D. Zhu, K. de Groot, Three-dimensional nano-HAp/collagen matrix loading with osteogenic cells in organ culture. J. Biomed. Mater. Res. A **44**(4), 407–415 (2004)
88. J.D. Hartgerink, E. Beniash, S.I. Stupp, Self-assembly and mineralization of peptide-amphiphile nanofibers. Science **294**, 1684–1688 (2001)
89. D.A. Wahl, J.T. Czernuszka, Collagen-hydroxyapatite composites for hard tissue repair. Eur. Cells Mater. **11**, 43–56 (2006)
90. S.A. Catledge, W.C. Clem, N. Shrikishen, S. Chowdhury, A.V. Stanishevsky, M. Koopman, Y.K. Vohra, An electrospun triphasic nanofibrous scaffold for bone tissue engineering. Biomed. Mater. **2**(2), 142–150 (2007)
91. D.W. Green, Tissue bionics: examples in biomimetic tissue engineering. Biomed. Mater. **3**, 034010 (2008)
92. A. Ficai, E. Andronescu, G. Voicu, D. Manzu, M. Ficai, Layer by layer deposition of hydroxyapatite onto the collagen matrix. Mater. Sci. Eng. C **29**, 2217–2220 (2009)
93. Y. Han, S. Li, X. Wang, X. Chen, Synthesis and sintering of nanocrystalline hydroxyapatite powders by citric acid sol–gel combustion method. Mater. Res. Bull. **39**, 25–32 (2004)
94. A. Tampieri, G. Celotti, El. Landi, M. Sandri, N. Roveri, G. Falini, Biologically inspired synthesis of bone-like composite: self-assembled collagen fibers/hydroxyapatite nanocrystals. J. Biomed. Mater. Res. A, **67A**(2), 618–625 (2003)

95. Ch. Jäger, T. Welzel, W. Meyer-Zaika, M. Epple, A solid-state NMR investigation of the structure of nanocrystalline hydroxyapatite. Magn. Reson. Chem. **44**(6), 573–580 (2006)
96. G.E. Poinern, R.K. Brundavanam, N. Mondinos, Z.-T. Jiang, Synthesis and characterisation of nanohydroxyapatite using an ultrasound assisted method. Ultrason. Sonochem. **16**, 469–474 (2009)
97. R. Khanna, K.S. Katti, D.R. Katti, Nanomechanics of surface modified nanohydroxyapatite particulates used in biomaterials. J. Eng. Mech ASCE **135**(5), 468–478 (2009)
98. A. Seilacher, Arbeitskonzept zur Konstruktionsmorphologie (Concept for structure-morphology). Lethaia **3**, 393–396 (1970), in German
99. S.J. Gould, R.C. Lewontin, The spandrels of San Marco and the Panglossian paradigm: a critique of the adaptionist program. Proc. R. Soc. Lond. B **205**(1161), 581–598 (1979)
100. Y. Kolodny, B. Luz, M. Sander, W.A. Clemens, Dinosaur bones: fossils or pseudomorphs? the pitfalls of physiology reconstruction from apatitic fossils. Palaeogeogr. Palaeoclimatol. Palaeoecol. **126**, 161–71 (1996)
101. N.K. Mathur, C.K. Narang, Chitin and chitosan, versatile polysaccharides from marine animals. J. Chem. Educ. **67**, 938–942 (1990)
102. S. Weiner, L. Addadi, H.D. Wagner, Materials design in biology. Mater. Sci. Eng. C **11**, 1–8 (2000)
103. S. Lees, Elastic properties and measurement techniques of hard tissues, in *Handbook of Elastic Properties of Solids, Liquids, and Gases*, Volume III: Elastic Properties of Solids, Chapter 7, ed. by M. Levy, H. Bass, R. Stern (Academic, New York, 2001), pp. 147–181
104. H.C.W. Skinner, A.H. Jahren, Biomineralization, in *Treatise on Geochemistry*, Volume 8: Biogeochemistry, Chapter 4, ed. by W.H. Schlesinger (Elsevier, Amsterdam, The Netherlands, 2003), pp. 117–184
105. I.C. Gebeshuber, H. Stachelberger, B.A. Ganji, D.C. Fu, J. Yunas, B.Y. Majlis, Exploring the innovational potential of biomimetics for novel 3D MEMS. Adv. Mater. Res. **74**, 265–268 (2009)

Chapter 6
Bioinspired Cellular Structures: Additive Manufacturing and Mechanical Properties

J. Stampfl, H.E. Pettermann, and R. Liska

Abstract Biological materials (e.g., wood, trabecular bone, marine skeletons) rely heavily on the use of cellular architecture, which provides several advantages. (1) The resulting structures can bear the variety of "real life" load spectra using a minimum of a given bulk material, featuring engineering lightweight design principles. (2) The inside of the structures is accessible to body fluids which deliver the required nutrients. (3) Furthermore, cellular architectures can grow organically by adding or removing individual struts or by changing the shape of the constituting elements. All these facts make the use of cellular architectures a reasonable choice for nature. Using additive manufacturing technologies (AMT), it is now possible to fabricate such structures for applications in engineering and biomedicine. In this chapter, we present methods that allow the 3D computational analysis of the mechanical properties of cellular structures with open porosity. Various different cellular architectures including disorder are studied. In order to quantify the influence of architecture, the apparent density is always kept constant. Furthermore, it is shown that how new advanced photopolymers can be used to tailor the mechanical and functional properties of the fabricated structures.

6.1 Introduction

Cellular structures are one of the basic design elements of many biological materials (e.g., wood, trabecular bone, marine skeletons) [1–4]. Cellular design offers several benefits: the resulting structures can bear the endurable mechanical loads using a minimum of a given bulk material, thus enabling the use of lightweight design principles. The interior of the structures is accessible to body fluids which deliver

J. Stampfl (✉)
Institute of Materials Science and Technology, Vienna University of Technology (TU Wien), Favoritenstrae 9, 1040 Vienna, Austria
e-mail: jstampfl@pop.tuwien.ac.at

the required nutrients. Furthermore, cellular architectures can grow organically by adding or removing individual struts or by changing the shape of the constituting elements. All these facts make the use of cellular architectures a reasonable choice for nature. Using additive manufacturing technologies (AMT), it is now possible to fabricate such cellular structures with almost arbitrary complexity. Direct [5–9] and indirect [10, 11] AMT have been used to fabricate cellular structures, whose most appealing application is the use as scaffold for biomedical engineering. Cellular structures made by AMT are depicted in Fig. 6.1.

The final goal for applications in biomedicine is the fabrication of structures with defined structural as well as functional properties [12]. Biologically inspired materials offer a large potential to tailor these properties. Regarding structural properties, biological cellular structures exhibit a significantly more irregular architecture than cellular engineering structures (e.g., truss structures). The role of irregularity in biological materials is not evident. Irregularity in natural biomaterials cannot be justified by a lack of "manufacturing skills" since many biomaterials (e.g., the skeleton of radiolaria, Fig. 6.2b, or the skeleton of deep-sea sponge *Euplectella aspergillum* [1]) exhibit a very regular and well-defined microarchitecture. Nevertheless, the microarchitecture of most cellular biomaterials (e.g., cancellous bone,

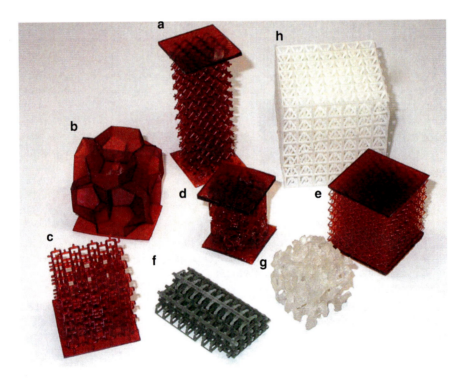

Fig. 6.1 Sample structures fabricated by various additive manufacturing processes. (**a–e**) fabricated by digital light processing, (**f**) manufactured by wax printing, (**g, h**) fabricated by selective laser sintering [13]

6 Bioinspired Cellular Structures: Additive Manufacturing and Mechanical Properties

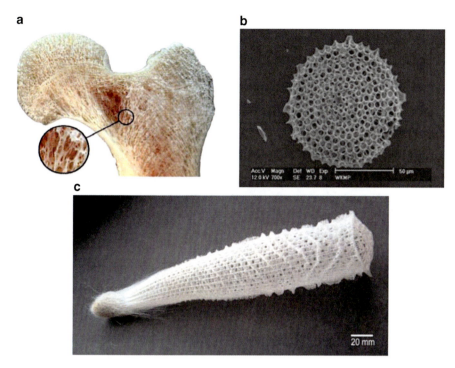

Fig. 6.2 Biomaterials like trabecular bone (**a**), diatoms (**b**) or *Euplectella aspergillum* (**c**) are based on cellular architecture [14]

see Fig. 6.2a, which shows a cross section through a human femur) is not completely regular.

On the level of microarchitecture, cancellous bone is made up of an interconnected network of rods or plates (*trabeculae*). The trabeculae are mainly oriented along the trajectories of applied mechanical stress. The trabeculae along the stress direction are connected by another class of trabeculae which is oriented perpendicular to the stress direction to mechanically stabilize the cellular network. As can be seen from Fig. 6.2a, there is a significant degree of irregularity on the level of microarchitecture.

6.2 Fabrication of Bioinspired Cellular Solids Using Lithography-Based Additive Manufacturing

Besides lithography-based systems, which are described in the following, several additional processes have gained a lot of attention and are available on a commercial basis. Fused deposition modeling (FDM) [15], selective sintering (SLS) [16], and

3D printing [6, 17] have to be mentioned due to their large installation base. Most of these processes are capable of fabricating cellular structures. Nevertheless, lithography-based additive manufacturing systems offer several advantages: on one hand, the feature resolution is significantly better than with all other available technologies. On the other hand, the functional properties of the obtained photopolymer can be tuned quite easily by varying the cross-link density of the polymer and by changing the utilized reactive diluents and cross-linkers (see Sect. 6.3).

6.2.1 Laser-Based Stereolithography

With a market share of 42% in the field of service bureaus, laser-based stereolithography is still dominating the high-end market of AMT. The basic working principle of stereolithography (SLA) has not changed since its invention in the 1980s of the twentieth century. The light source of the system is made up of a UV-laser, typically a solid state Nd-YAG-laser. Using a galvanoscanner, the light beam is deflected according to the geometry of the CAD file. Where the laser beam hits the resin surface, polymerization takes place and the liquid resin solidifies. After one layer has been completed, the part is lowered into the vat and the coating mechanism deposits a fresh layer of resin on top of the part. The process is repeated until all layers have been built.

If the part geometry contains severe undercuts, support structures are necessary since the liquid resin is not capable of stabilizing geometries which have no connection to the main part. The generation of support structures is achieved automatically by modern AMT software. Problems might arise when the supports, which have to be removed mechanically, have to be detached from delicate features.

Commercial SLA systems typically use layer thicknesses of 50–100 μm. Due to the high scan speeds of 500–1,000 mm/s and the large possible build-volumes, SLA is mainly used for high-throughput jobs in service bureaus. With an appropriate optical setup, the feature resolution and minimal layer thickness can be improved significantly and using micro-SLA-systems lateral resolutions of down to 5 μm have been achieved [18]. Sample structures fabricated by this process are depicted in Fig. 6.3.

6.2.2 Dynamic Mask-Based Stereolithography

In order to overcome the price limits given by the fairly expensive UV-laser-systems, lithography-based processes using visible light in combination with a dynamic mask have been developed. The main benefit of these systems is the fact that they can rely on components which are widely used in consumer products (e.g., video beamers).

Dynamic mask-based stereolithography (DMS) systems offer distinct benefits and drawbacks compared to laser-based stereolithography: Due to the fact that the whole layer is exposed in one step, the fabrication speed is faster. Typical exposure times for one layer range from 3–12 s. Since most DMS systems use a high-pressure

6 Bioinspired Cellular Structures: Additive Manufacturing and Mechanical Properties

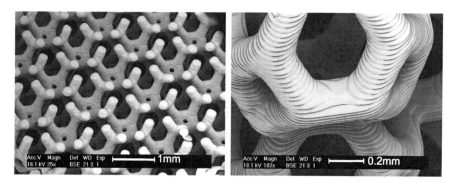

Fig. 6.3 Cellular structures fabricated by microstereolithography [18]

mercury lamp instead of a UV-laser, the cost for replacing and maintaining the light source are fairly moderate. The use of high-pressure mercury lamps and micromirror arrays necessitates the use of visible light for the polymerization of the resin. Most micromirror arrays cannot be used in combination with UV, although recent developments might change this restriction. The light intensity of the available light sources is smaller than in the case of UV-lasers. Both limitations lead to the fact that well-established and widely used SLA resins cannot be used in combination with DMS.

A further drawback is related to the number of pixels which are available on the chip. At a given build volume, the achievable pixel resolution is given by the number of mirrors on the chip. For a system with $1,400 \times 1,050$ pixels and a targeted resolution of $50\,\mu m$, the build size is 70×50 mm. DMS systems are therefore mostly used for the fabrication of delicate and small parts. If the required build size is very large, the possible resolution will be too low.

A typical schematic setup for a DMS system is given in Fig. 6.4. The light source emits light which is selectively deflected by the micromirrors on the DLP chip (digital light processing). An appropriate lens focuses the image on the bottom surface of the transparent vat. White pixels lead to a solidification of the photosensitive resin. In regions of black pixels, the resin stays liquid. After solidification has taken place, the build platform is lifted in z-direction, according to the layer thickness. Liquid resin flows into the now empty space and the process is repeated. Exposure from underneath offers several advantages compared to the exposure from top (which is used in traditional laser-based processes). The main advantage from an economical point of view is the fact that significantly less resin is needed to run the bottom-exposure process compared to the top-exposure process. In Fig. 6.4, the necessary resin level is only a couple of millimeters in height, independently of the height of the part. In a top-exposure setup, the resin level has to be high enough, as to accommodate the complete final part.

The main advantage from an engineering point of view is the possibility to expose very thin layers with little processing effort. The layer thickness can easily be adjusted by defining the gap between the previous layer and the transparent bottom of the vat. No special coating mechanism is necessary, and even small layer thicknesses down to 10–$20\,\mu m$ are possible.

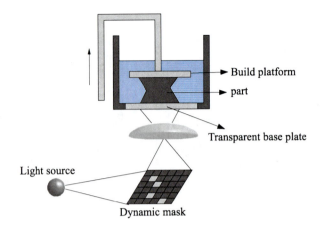

Fig. 6.4 Working principle of dynamic mask-based stereolithography

One drawback of the exposure from the bottom are issues with adhesion between polymerized resin and the transparent vat. The vat has to be coated with silicone and/or a transparent teflon layer to prevent large pull-off forces when the polymerized body is detached from the vat. If the resin contains initiator molecules or monomers with low molecular weight, these compounds tend to diffuse into the vat, leading to unwanted polymerization inside the separation layer.

Due to the fact that the whole layer is exposed in one step, DMS systems work fairly fast. Typically, vertical build speeds between 10 and 20 mm/h are achieved. Due to the moderate hardware cost, the ability to fabricate complex parts with excellent feature resolution, DMS systems are widely used for applications in jewellry (master patterns for investment casting) and the fabrication of hearing aids.

6.2.3 Inkjet-Based Systems

In order to combine the benefits of lithographic methods (high feature resolution and good surface quality) with the advantages of inkjet-based processes (high build speed and large build volume), two companies (Objet, 3D-Systems) have developed machines which are capable of processing liquid photopolymers using an inkjet head. In commercially available systems, an inkjet head with several hundred nozzles is moved along the x-axis, and during this motion the nozzles eject little droplets of photopolymer. After one layer is deposited, a UV-lamp flash-cures the fresh layer and the process is repeated.

The inkjet head can deposit two types of material: The build material and a support material. The support material, which can be removed mechanically after the build process has been completed, is necessary to serve as base for the build material in regions with overhangs. In contrast to conventional stereolithography, where light-weight supports are only required in areas with severe overhangs, inkjet-based processes require a completely dense support structure. The overall amount of material (build plus support) is therefore identical to the build volume of the part. Due to the use of multiple inkjet heads, it is possible to build multimaterial structures, which is impossible in the case of SLA and DMS systems. Since the inkjet-head can be moved over arbitrary distances, large parts with good feature resolution can be built. Nevertheless there are limitations, especially regarding the allowed viscosity of the photopolymer. The viscosity has to be kept within a narrow range to be processable inside the inkjet head.

6.2.4 Two-Photon Polymerization

Two photon polymerization (2PP) [19,20] offers two distinct benefits with respect to other AMT: (1) The achievable feature resolution is about one order of magnitude better than with other additive manufacturing methods. The minimum achievable wall thickness is currently around 100 nm [21]. (2) Furthermore, it is possible to directly write inside a given volume. In contrast, all other AMTs work by shaping individual 2D layers and stacking up these layers to fabricate a 3D model. Due to this additive stacking process, it is not possible to embed existing components into a part made by traditional AMT. In contrast, 2PP is capable of writing "around" pre-embedded components. Despite these distinct advantages of 2PP, there are currently no commercial 2PP applications, mainly due to the low writing speed of 2PP systems and due to the complexity of the required lasers. With the development of more efficient initiators [22] and more powerful femtosecond lasers, this situation is expected to improve in the near future.

Besides the fabrication of high-resolution structures (see cellular structures in Fig. 6.5), 2PP is especially suitable for applications where existing components have to be embedded inside the written structure. For instance, 2PP can be used to connect optoelectronic components (laser diodes and photo diodes) by an optical waveguide [23].

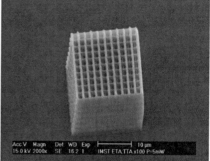

Fig. 6.5 Cellular structures made by two-photon polymerization

6.3 Photopolymers for Additive Manufacturing Technologies

6.3.1 Principles of Photopolymerization

The process of photopolymerization has been introduced to industry more than half a decade ago. Coatings and printing inks for wood, paper, card board, plastics, and metal were among the first applications of this promising technology. Compared to the classical hardening of coatings by evaporation of the solvent and/or a thermal curing step, UV formulations can be cured within a fraction of a second. Therefore, photopolymerization is a continuously growing technology, finding new applications on a regular basis. This includes dental filling materials, reprography (photoresists, printing plates, integrated circuits), holographic recordings, nanoscale micromechanics, or stereolithography.

One can distinguish between two different curing mechanism: radical and cationic. While radical curing is the predominant technique in industry, the importance of cationic systems is continuously growing.

Typical monomers for radical polymerization are based on acrylates or methacrylates that can be cured by UV–Vis sensitive compounds [24]. Such photoinitiators form a large amount of initiating radicals when irradiated with light of particular wavelength. Those radicals start the free radical polymerization chain reaction. The liquid formulation is converted into a solid and usually cross-linked material.

In the first step (*initiation*), see (1) in Fig. 6.6, the PI dissociates into radicals ($R_i^\bullet$) due to absorption of light. Those radicals add onto monomer molecules to start the chain reaction, see (2) in Fig. 6.6. During the chain growth reaction (*propagation*), the addition of other monomer molecules occurs, whereas chain *termination* takes place because of recombination or disproportionation (see (3) in Fig. 6.6). The generation of radicals in the first step can either happen through photofragmentation

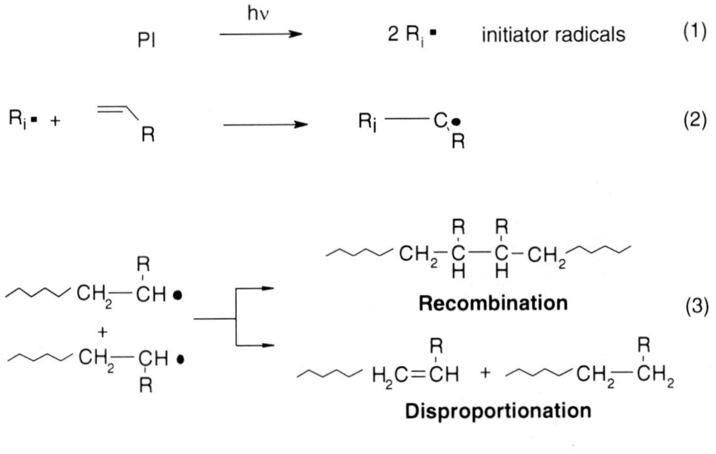

Fig. 6.6 Principle of photopolymerization

6 Bioinspired Cellular Structures: Additive Manufacturing and Mechanical Properties

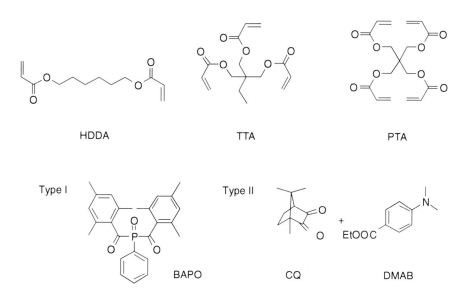

Fig. 6.7 Common monomers for photopolymerization and photoinitiators based on Type I and Type II mechanism

as a result of α-cleavage (Type I) or through hydrogen abstraction or electron transfer (Type II) in a bimolecular mechanism. Typical examples (Fig. 6.7) for such photoinitiators are bisacylphosphine oxides (BAPO) or the combination of camphorquinone (CQ) with an aromatic amine (DMAB).

Sometimes, polymerization of the highly reactive monomers can even start during storage due to thermal stress. To avoid this, it is necessary to use inhibitors that scavenge accidentally formed radicals before they start the polymerization. Oxygen can also cause the inhibition of the radical polymerization, but during photocuring it is an undesired process leading to sticky surfaces. Therefore, care has to be taken as to reduce the amount of oxygen present, since lower double bond conversion can result in migratables.

Acrylates or methacrylates are widely used for the radical photopolymerization due to their high reactivity [24]. Figure 6.7 displays common multifunctional monomers such as 1,6-hexanediol diacrylate (HDDA), trimethylolpropane triacrylate (TTA), and pentaerythritol tetraacrylate (PTA). Such monomers are usually applied as reactive diluents, for high molecular monomers based on polyester-, urethane-, or epoxy-(meth)acrylates that are generally responsible for the overall performance of a formulation.

For cationic curing, mainly epoxy-based monomers e.g., ECC or BDG are used, (Fig. 6.8). Less frequently also vinylethers are used. Generally, the cationic polymerization process is more temperature sensitive. Heating after irradiation can increase the speed of polymerization and lead to a reduction of the postcuring time. Chain transfer agents based on alcohols like CDM are applied to tune the

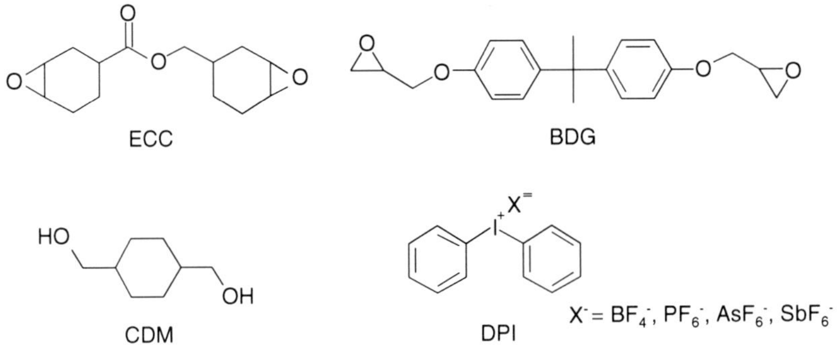

Fig. 6.8 Typical epoxy monomers (*top row*) and chain transfer agent and PIs (*bottom row*; left to right) based on cationic polymerization

polymerization network. Onium salts such as DPI are most frequently used as photoinitiator, but these molecules suffer from poor absorption in the UV-A and UV-B. Therefore, sensitizers such as isopropyl thioxanthone have to be applied that extend the wavelength of sensitivity up to the visible range of the spectrum.

6.3.2 Radical and Cationic Systems in Lithography-Based AMT

In early days of the development of UV curable resin formulations for stereolithography, systems were based on radical or cationic polymerization [25, 26]. As there was already a large variety of monomers commercially available from the coating industry, formulators could easily select the desired components to optimize the curing speed and tune the final mechanical properties over a wide range. Both photopolymerizable systems have their own pros and cons. Radical polymerization proceeds very fast, but relatively high shrinkage and oxygen inhibition are the crucial disadvantages. In cationic polymerization, these drawback are avoided, but poor reactivity, sensitivity to moisture, and high water uptake of the polymer are adverse effects. To combine the advantages of both systems, several commercially available formulations nowadays include both concepts, these formulations being called hybrid formulations. In this case, careful adjustment of the amounts of radical and cationic initiator based on their absorption behavior is essential to ensure photochemical excitation of both molecules.

It is important that the absorption behavior of the photoinitiators match the emission characteristics of the lamp or laser used. For systems working only in the visible range of the spectrum, specialized photoinitiators such as those shown in Fig. 6.7 have to be used. Most other commercially used photoinitiators such as Darocure 1173, Irgacure 2959, Irgacure 369 (all Ciba SC) only absorb in the UV and therefore can only be used in machines with broadband irradiation of, e.g., mercury arc lamps or by laser excitation, e.g., at 355 nm.

It has to be noted that the two-photon polymerization is usually carried out with similar photoinitiators, but due to the low two-photon absorption cross section, inefficient curing has to be accepted. Therefore, several research groups have focused recently on the development of specialized photoinitiators, which are mainly based on a planar conjugated chromophor with an acceptor group in the middle of the molecule and with terminal donor groups [27, 28].

Beside the photopolymerizable monomers and photoinitiators, additives and fillers are essential. For example, antifoaming agents are required to avoid air bubbles. Furthermore, titanium dioxide or surface-modified aerosils are used to tune the final mechanical properties. Polysiloxanes are added for optimum dispersion of these fillers.

6.3.3 Biomimetic, Biocompatible, and Biodegradable Formulations

As life expectancy increases with advances in medical care during the last decades, tissue engineering of bone, blood vessels, skin, and other organs has gained significant interest. In the last few years, additive manufacturing has been recognized as an ideal tool to print individual 3D cellular scaffolds. Dealing with a complex and sensitive biological system such as the human body, the development of materials for tissue engineering is extremely challenging and the requirements are manifold. Biocompatibility, from a chemical, structural, and mechanical point of view, is the main issue. Degradation times must match the healing and regeneration process of the replaced tissue and lead to nontoxic degradation products.

Classical biodegradable materials are based on poly(α-esters) such as poly(lactic acid) (PLA), poly(glycolic acid) (PGA), or copolymers thereof (Fig. 6.9). These materials are already FDA approved and can be found in clinical applications. Unfortunately, they can only be processed by classical techniques for thermoplasts, like extrusion or injection molding. These and alternative methods such as fiber spinning and bonding, solvent casting, particulate leaching, or gas foaming possess some inherent limitations in the construction of internal channels and the capability to precisely control pore size, pore geometry, pore interconnectivity, and spatial distribution of pores within the scaffold.

To use the inherent advantages of AM, photopolymers have to be used. In the field of biocompatible photopolymers, (meth)acrylated PEG chains of various length (e.g., PEG-DM) have been used for many applications, as uncontrolled cell adhesion can be avoided. Examples for applications include the encapsulation of chondrocytes for cartilage regeneration, osteoblasts for bone tissue engineering, vascular smooth muscle cells, and mesenchymal stem cells. Anseth et al. [29] were among the first to synthesize degradable methacrylate endcapped oligoesters based on polylactic acid (PEG-PLA-DM), which enabled processing by light-induced curing and opened the door to stereolithographic tissue engineering.

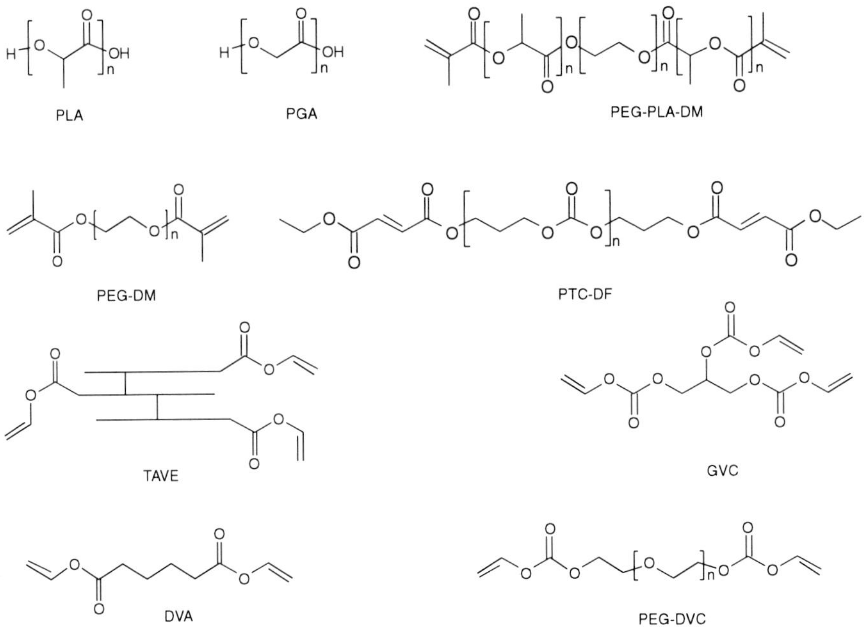

Fig. 6.9 Biocompatible monomers and polymers

Unfortunately, polylactic acid block copolymers suffer from the well-known hydrolytic bulk degradation mechanism which causes quite fast loss of the mechanical properties. The reason for that is the formation of acidic degradation products within the scaffold that cannot migrate and lead to an autocatalytic degradation process. To have a more controlled enzymatic degradation and to force cell adhesion and proliferation, methacrylated gelatin derivatives have been prepared [30]. Very recently, Feijen and Grijpma [31] used the concept of end-functionalized poly(trimethylene carbonate) PTC-DF that avoid the formation of larger amounts of acidic degradation products. They have also used fumaric acid as an alternative concept of photopolymerizable groups. Therefore, they are able to avoid (meth)acrylates that always have some inherent skin irritancy or sometimes cytotoxicity, but they have to accept a significantly lower photoreactivity. Further drawbacks of (meth)acrylates as photopolymerizable biodegradable material is the rather long kinetic chain length, which leads to polymer backbones with molecular weight of significantly more than 50,000 g/mol.

As the molecular weight limit in the transport within the human body is in the order of a few thousand g/mol, this polyacrylic acid degradation product (Fig. 6.10) cannot be excreted. Also the cross-linking of the polyacrylic acid with, e.g., Ca^{2+} ions, hinders the transport within the human body. Additionally, high local concentration of acids might lead to inflammation reactions or, in the worst case, necrosis and fast unwanted bulk degradation.

6 Bioinspired Cellular Structures: Additive Manufacturing and Mechanical Properties

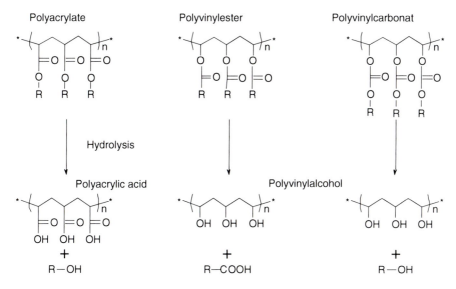

Fig. 6.10 Degradation of polymers

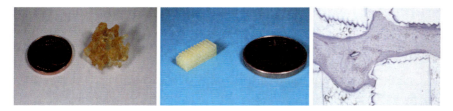

Fig. 6.11 3D-cellular structure printed by AM out of a vinylcarbonat polymer (*left*) and first results of in vivo studies of vinylesters (*right*) [34]

An alternative polymer backbone is based on polyvinyl alcohol, which is FDA approved and frequently used in medical applications. Well-known polymers from vinylesters can be considered as suitable starting product, but photopolymerization is rarely described until now [32]. In this case, low molecular acidic degradation products are formed (Fig. 6.10). These degradation products can easily be transported and metabolized within the human body. To completely avoid organic acids as degradation products, vinylcarbonates such as PEG-DVC or GVC were recently introduced [33]. This class of monomers not only has an extraordinary low cytotoxicity, but excellent storage stability and sufficient photoreactivity have also been determined. By proper tuning of the components, mechanical properties of PLA can be surpassed and degradation behavior can be easily tuned within a broad range. Preliminary in vivo experiments showed promising results in the New Zealand white rabbit model (Fig. 6.11).

6.4 Mechanical Properties: Modeling and Simulation

The mechanical properties of cellular materials are fundamentally different from conventional, dense materials. Thus, they cannot be modeled sufficiently well by common, standard constitutive material laws. Due to the highly porous structure, the behavior is not only determined by the building materials, but is also significantly governed by the structural architecture. Moreover, most of the cellular materials – both natural and man-made ones – exhibit direction-dependent properties.

The aim of mathematical modeling is to derive relations between the structural architecture and the behavior of cellular materials [2, 35]. On one hand, such investigations can gain better understanding of biological materials which are commonly regarded as being "optimized." On the other hand, for the improvement of the performance of technical materials, modeling and simulation are the keys. They allow for designing the structural architecture, as to meet a specified property profile and, at the same time, to keep the weight to a minimum.

These investigations are preferably conducted on a variety of generic structures. Their architectures can be selected such that specific mechanisms of deformation will be activated. By means of the finite element method (or other numerical and analytical tools), the response to loads and load sequences can be predicted.

6.4.1 Linear Elastic Behavior

First, the elastic behavior will be discussed [36, 37]. Various generic architectures, all sharing the same specific weight, exhibit a wide range of properties. Each structure shows its own characteristic elasticity profile with respect to the direction dependence. Moreover, comparison among the structures reveals a high variability in the magnitude of the stiffnesses. In a systematic study, the density of all structures (i.e., the volume share of the base material) can be varied and the influence of the density on the mechanical properties can be assessed for each structure. It turns out that there exists an exponential relation [2] between the density and the Young's modulus characterizing the response to uniaxial stresses, reading as:

$$E^* = E_b \, \rho^\beta, \tag{6.1}$$

where E^* and E_b are the Young's moduli of the cellular material and of the base material, respectively, ρ is the relative density of the cellular material, and β is the density exponent. The latter one takes values between 1 and 3, depending upon which deformation mechanisms are activated upon loading. Plate bending of the members in cellular materials is indicated by a value of 3, whereas normal and shear deformation are associated with an exponent of one. Such pure deformation mechanisms only occur in very particular cases. Typically the mechanisms are superimposed, even for quite simple architectures under basic loading. For example, a simple cubic arrangement of struts shows an exponent of 1 in principal direction and a value of almost 2 in diagonal direction. Consequently, the density exponent is not a fixed number to be valid for a particular cellular structure, but it is very much dependent on the direction of loading [37]. Last but not least, the density exponent can be taken as measure which identifies the active mechanisms of deformation in a loaded cellular material.

6.4.2 Nonlinear Response

If the load imposed on the cellular material is increased, the limit of the elastic range will be reached. Then various types of nonlinearities may occur in the base material, like nonlinear elasticity or elastoplasticity. Additionally, viscous effects can play a role in materials such as bone tissue, polymers, and metals at elevated temperatures. All these types of material behavior can be incorporated into the model, as to predict the response of cellular materials [38, 39].

One of the most important effects beyond elasticity, however, is local loss of stability. In such an event, the cell walls or struts start to buckle or kink which leads to a drastic loss of their stiffness and the neighboring cell walls have to carry additional loads. If they are not capable of bearing those, they will also buckle and a chain reaction will be triggered giving rise to a catastrophic localized failure which will be located at one plane in the cellular material. Modeling and simulation can be used to investigate the susceptibility to failure with respect to the cellular architecture and the loading scenario. Along the same line, limit loads and the post-buckling behavior can be predicted [39].

Most of the biological cellular materials possess some degree of structural disorder and imperfection. These effects are treated by starting with perfect generic architectures and introducing statistical disorder in a controlled way. In parallel, the influence of defects in cellular materials is studied. Therefore, cell walls and struts can be removed in a randomly uniform manner and in clustered configurations of different sizes. This way, missing or broken connections are modeled, as well as vacancies and pores [40,41]. It is shown that with increasing disorder, the maximum values of the properties decrease for certain directions. In general, however, a more balanced property profile is achieved. In particular, structures whose degree of disorder exceeds a certain level are less prone to deformation localization and catastrophic failure. Also, the defect sensitivity is decreased for disordered architectures.

In addition to the computational studies of the behavior, experimental investigations are carried out. This way, the predictive quality of the simulations are verified with respect to elastic and strength properties, as well as with respect to the failure characteristic including deformation localization.

6.4.3 Sample Size and Effective Behavior

Typically, theoretical investigations are performed on various length scales simultaneously. On the small scale, the cell geometry and their walls and struts are described with respect to their material behavior and their local geometrical features

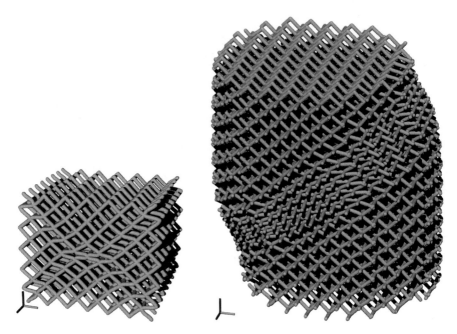

Fig. 6.12 FEM predictions of the deformation localization of samples with different size but the same cellular architecture [13]

including the arrangement. Many of such members form a diverse cellular structure, which now – on the large scale – can be regarded as a material again. Properties can be assigned which characterize their global or effective or "homogenized" behavior. For such considerations to hold true, the cellular structure must be relatively large and their local features should not vary (in a statistical sense). As a rule of thumb, the linear dimension of the structure must be at least one order of magnitude larger than the local characteristic cell size.

If this "separation of length scales" is not fulfilled, the interpretation of results must be handled with great care. This applies not only to modeling and simulation, but also to experimental testing. Some effects may be attributed to the global material behavior, and some others may rather be a feature of a small structure. Influences originating from the perimeter of the structure, such as load introduction or force-free faces, can govern the behavior or trigger mechanisms not owing to the behavior of a cellular material. Typical examples are cancellous bone in the proximal femur or specimens of limited size for experimental testing. Figure 6.12 shows the predicted deformation localizations of two samples with different size, but identical cellular architecture. Localization in the left sample is triggered at the corners and, thus, reflects the behavior of the sample rather than that of the cellular material. In contrast, the central part of the right sample reflects the behavior of the cellular material which does not show confined localization.

6.5 Conclusion

Within this chapter, it has been shown that lithography-based AMT is a suitable tool to print arbitrarily complex structures from a CAD file out of photopolymers. To broaden the range of application, different concepts were shown to structure biocompatible or biodegradable materials. Beside inkjet printing, dynamic-mask-based stereolithography, and microstereolithography, two photon polymerization is a promising real 3D-writing tool to build parts with feature resolution well below 1 μm. AMT is especially useful for fabricating cellular structures which mimic the cellular architectures, which can be found in nature. Using finite element modeling, it is possible to predict the mechanical behavior of such cellular structures. Of initial interest are the linear elastic properties. By further extending the FEM models into the nonlinear regime, predictions about the failure behavior of cellular structures can be made.

References

1. J. Aizenberg, M. Thanawala, V. Sundar, J. Weaver, D. Morse, P. Fratzl, Materials science: skeleton of Euplectella sp.: structural hierarchy from the nanoscale to the macroscale. Science **309**, 275–278 (2005)
2. L.J. Gibson, M.F. Ashby, Cellular Solids, 2nd edn. (Cambridge University Press, Cambridge, 1997)
3. P. Fratzl, R. Weinkamer, Nature's hierarchical materials. Prog. Mater. Sci. **52**(8), 1263–1334 (2007)
4. M.A. Meyers, P.-Y. Chen, A.Y.-M. Lin, Y. Seki, Biological materials: structure and mechanical properties. Prog. Mater. Sci. **53**(1), 1–206 (2008)
5. C.X.F. Lam, X.M. Mo, S.H. Teoh, D.W. Hutmacher, Scaffold development using 3D printing with starch powder. Mater. Sci. Eng. C **20**, 49–56 (2002)
6. B. Leukers, H. Gulkan, S. Irsen, S. Milz, C. Tille, M. Schieker, H. Seitz, Hydroxyapatite scaffolds for bone tissue engineering made by 3D printing. J. Mater. Sci. Mater. Med. **16**(12), 1121–1124 (2005)
7. D.W. Hutmacher, Scaffold design and fabrication technologies for engineering tissues- state-of-the-art and future perspectives. J. Biomater. Sci. Polym. Ed. **12**(1), 107–124 (2001)
8. R. Bibb, G. Sisias, Bone structure models using stereolithography: a technical note. Rapid Prototyping J. **8**(1), 25–29 (2002)
9. A. Woesz, J. Stampfl, P. Fratzl, Cellular solids beyond the apparent density – an experimental assessment of mechanical properties. Adv. Eng. Mater. **6**(3), 134–138 (2004)
10. I. Manjubala, A. Woesz, C. Pilz, M. Rumpler, N. Fratzl-Zelman, P. Roschger, J. Stampfl, P. Fratzl, Biomimetic mineral-organic composite scaffolds with controlled internal architecture. J. Mater. Sci. Mater. Med. **16**, 1111–1119 (2005)
11. A. Woesz, M. Rumpler, J. Stampfl, F. Varga, N. Fratzl-Zelman, P. Roschger, K. Klaushofer, P. Fratzl, Towards bone replacment materials from calcium phosphates via rapid prototyping and ceramic gelcasting. Mater. Sci. Eng. C **25**(2), 181–186 (2005)
12. E.L. Hedberg, C.K. Shih, J.J. Lemoine, M.D. Timmer, M.A. Liebschner, J.A. Jansen, A.G. Mikos, In vitro degradation of porous poly(propylene fumarate)/poly(DL-lactic-co- glycolic acid) composite scaffolds. Biomaterials **26**(16), 3215–3225 (2005)

13. M.H. Luxner, *Modeling and Simulation of Highly Porous Open Cell Structures – Elasto-Plasticity and Localization Versus Disorder and Defects*. PhD thesis, Technische Universität Wien, 2006
14. J. Stampfl, M.H. Luxner, H.E. Pettermann, Zellulare Werkstoffe mit frei wählbarer Zellgeometrie – Herstellung, Modellierung der mechanischen Eigenschaften und Anwendungen. Lignovisionen **14**, 55–60 (2006)
15. I. Zein, D.W. Hutmacher, K.C. Tan, S.H. Teoh, Fused deposition modeling of novel scaffold architectures for tissue engineering applications. Biomaterials **23**(4), 1169–1185 (2002)
16. M. Agarwala, D. Bourell, J. Beaman, H. Marcus, J. Barlow, Direct selective laser sintering of metals. Rapid Prototyping J. **1**(1), 26–36 (1995)
17. J. Moon, J.E. Grau, V. Knezevic, M.J. Cima, E.M. Sachs, Ink-jet printing of binders for ceramic components. J. Am. Ceramic Soc. **85**(4), 755–762 (2002)
18. J. Stampfl, S. Baudis, C. Heller, R. Liska, A. Neumeister, R. Kling, A. Ostendorf, M Spitzbart, Photopolymers with tunable mechanical properties processed by laser-based high-resolution stereolithography. J. Micromech. Microeng. **18**(12), 125014 (2008)
19. S. Passinger, M.S.M. Saifullah, C. Reinhardt, K.R.V. Subramanian, B.N. Chichkov, M.E. Welland, Direct 3D patterning of TiO2 using femtosecond laser pulses. Adv. Mater. **19**(9), 1218–1221 (2007)
20. S. Maruo, O. Nakamura, S. Kawata, Three-dimensional microfabrication with two-photon-absorbed photopolymerization. Opt. Lett. **22**(2), 132–134 (1997)
21. S.H. Park, T.W. Lim, D.Y. Yang, R.H. Kim, K.S. Lee, Improvement of spatial resolution in nano-stereolithography using radical quencher. Macromol. Res. **14**(5), 559–564 (2006)
22. C. Heller, N. Pucher, B. Seidl, L. Kuna, V. Satzinger, V. Schmidt, H. Lichtenegger, J. Stampfl, R. Liska, One- and two-photon activity of cross-conjugated photoinitiators with bathochromic shift. J. Polym. Sci. A Polym. Chem. **45**, 3280–3291 (2007)
23. J. Stampfl, R. Infuehr, S. Krivec, R. Liska, H. Lichtenegger, V. Satzinger, V. Schmidt, N. Matsko, W. Grogger, 3D-structuring of optical waveguides with two photon polymerization, in *Material Systems and Processes for Three-Dimensional Micro- and Nanoscale Fabrication and Lithography*, ed. by S.M. Kuebler, V.T. Milam, volume 1179E of *Mater. Res. Soc. Symp. Proc.*, Warrendale, PA, 2009, pp. 1179–BB01–07
24. K. Dietliker, *Chemistry and Technology of UV and EB Formulation for Coatings, Inks and Paints Vol. 3: Photoinitiators for Free Radical and Cationic Polymerisation* (SITA Technology Ltd., London, 1991)
25. J. Crivello, Radiation-curable cycloaliphatic epoxy compounds, uses thereof, and compositions containing them, 2000. US6075155A
26. S.K. Mirle, R.J. Kumpfmiller, Photosensitive composition useful in three-dimensional part-building and having improved photospeed, 1995. WO9513565A1
27. M. Rumi, S. Barlow, J. Wang, J.W. Perry, S.R. Marder, Two-photon absorbing materials and two-photon-induced chemistry. Adv. Polym. Sci. **213**, 1–95 (2008)
28. N. Pucher, A. Rosspeintner, V. Satzinger, V. Schmidt, G. Gescheidt, J. Stampfl, R. Liska. Structure-activity relationship in D-π-A-π-D-based photoinitiators for the two-photon-induced photopolymerization process. Macromolecules **42**, 6519–6528 (2009)
29. S. Lu, K.S. Anseth, Release behavior of high molecular weight solutes from poly(ethylene glycol)-based degradable network. Macromolecues **33**(7), 2509–2515 (2000)
30. M. Schuster, C. Turecek, G. Weigel, R. Saf, J. Stampfl, F. Varga, R. Liska, Gelatin-based photopolymers for bone replacement materials. J. Polym. Sci. A Polym. Chem. **47**, 7078–7089 (2009)
31. Q. Hou, D.W. Grijpma, J. Feijen, Creep-resistant elastomeric networks prepared by photocrosslinking fumaric acid monoethyl ester-functionalized poly(trimethylene carbonate) oligomers. Acta Biomater. **5**(5), 543–551 (2009)
32. H. Wei, T.Y. Lee, W. Miao, R. Fortenberry, D.H. Magers, S. Hait, A.C. Guymon, S.E. Joensson, C.E. Hoyle, Characterization and photopolymerization of divinyl fumarate. Macromolecules **40**, 6172–6180 (2007)

33. R. Liska, J. Stampfl, F. Varga, H. Gruber, S. Baudis, C. Heller, M. Schuster, H. Bergmeister, G. Weigel, C. Dworak, Composition that can be cured by polymerisation for the production of biodegradable, biocompatible crosslinkable polymers on the basis of polyvinyl alcohol, 2009. PCT Int. Appl. WO 2009065162 A2
34. C. Heller, M. Schwentenwein, G. Russmüller, F. Varga, J. Stampfl, R. Liska, Vinyl esters: low cytotoxicity monomers for the fabrication of biocompatible 3D scaffolds by lithography based additive manufacturing technologies. J. Polym. Sci. A Polym. Chem. **47**, 6941–6954 (2009)
35. T. Daxner, J.H. Böhm, M. Seitzberger, F.G. Rammerstorfer, Modeling of cellular metals, in *Handbook of Cellular Metals*, ed. by H.P. Degischer, B. Kriszt (Wiley-VCH, 2002), pp. 245–380
36. M.H. Luxner, J. Stampfl, H.E. Pettermann, Finite element modeling concepts and linear analyses of 3D regular open cell structures. J. Mater. Sci. **40**, 5859–5866, (2005)
37. M.H. Luxner, H.E. Pettermann, Modeling and simulation of highly porous open cell structures: Elasto-plasticity and localization versus disorder and defects, in *Proceedings of the IUTAM Symposium on Mechanical Properties of Cellular Materials*, IUTAM, ed. by H. Zhao, N.A. Fleck, (Springer-Verlag, Berlin, 2009), pp. 125–143
38. H.E. Pettermann, D. Garcia Vallejo, J. Stampfl, M.H. Luxner, J. Dominguez, Viscoelastic properties of open cell kelvin foams, in *Proceedings of XXII International Congress of Theoretical and Applied Mechanics (ICTAM)*, Adelaide, Australia, 25–29 August 2008
39. M.H. Luxner, J. Stampfl, H.E. Pettermann, Numerical simulations of 3D open cell structures – influence of structural irregularities on elasto-plasticity and deformation localization. Int. J. Solids Struct. **44**, 2990–3003 (2007)
40. M.H. Luxner, A. Woesz, J. Stampfl, P. Fratzl, H.E. Pettermann, A finite element study on the effects of disorder in cellular structures. Acta Biomater. **5**, 381–390 (2009)
41. M.H. Luxner, J. Stampfl, H.E. Pettermann, Nonlinear simulations on the interaction of disorder and defects in open cell structures. Comput. Mater. Sci. **47**, 418–428 (2009)

Part II
Form and Construction

Chapter 7
Biomimetics in Architecture [Architekturbionik]

Petra Gruber

Abstract This chapter presents an overview of this emerging field, investigating the overlaps between biology and architecture. A brief description of historic developments and classical approaches such as analogy research sets the stage for a strategic search for a new methodology of design. Different methods of biomimetic design are compared with regard to information transfer. Application fields in various scales and successful examples in architecture are presented to illustrate the use of basic biological phenomena such as emergence, differentiation, intelligence and interactivity in a technological as well as designed environment. Own projects as case studies in biomimetic design are described and compared to derive success factors for future projects.

7.1 Introduction

Biomimetics in Architecture [Architekturbionik] is an emerging field that is currently being defined and explored. The application of observations made in nature to architecture has always been a challenge for architects and designers. The strategic search for role models in nature is what discerns biomimetics from the ever-existing inspiration from nature. While bioinspiration may be limited to a morphological analogy, biomimetics makes use of functional analogies, processes, mechanisms, strategies or information derived from living organisms.

P. Gruber (✉)
transarch, office for biomimetics and transdisciplinary architecture, Zentagasse 38/1, 1050 Vienna, Austria
e-mail: peg@transarch.org
and
Institute for History of Art and Architecture, Building Archaeology and Preservation, Vienna University of Technology (TU Wien), Karlsplatz 13/251-1, 1040 Vienna, Austria
e-mail: petra.gruber@tuwien.ac.at

The term "biomimetics" appeared in the 1960ies, at about the same time as "bionics" was coined, that found its reinterpretation in the German "Bionik", the combination of the first and last syllables of the words "Biologie" (biology) and "Technik" (technology) [1]. Meanwhile Bionik/Biomimetics is widely accepted as an interdisciplinary science that delivers innovation in a wide range of application fields. Architecture, design and building are promising fields for biomimetic innovation.

7.2 History: Different Approaches

The overlaps between biology and architecture are manifold, and observation of interactions is based on the western culture of opposing nature and technology, often ignoring the dialectic perspective of interactiveness and mutual influence. As Portoghesi puts it: "Being an integral part of nature ourselves, we shall never be able to talk about it from the outside but only from the inside, uncertain whether to consider something created and produced by man as being 'outside' nature" [2].

Current developments as described in the following increasingly use a biological paradigm to overcome this separation.

Many movements in the history of architecture have taken their own approaches to nature, and famous architects such as Alvar Alto, Frank Lloyd Wright and Le Corbusier have laid emphasis on this connection. A chronological discussion of the history of architecture is beyond the scope of this chapter, but some important examples of the last century will illustrate the diversity.

A defined starting point in time for biomimetics in architecture cannot be stated. Inspiration from nature in architecture has existed at all times. The first human dwellings were natural shelters, and archetypes such as caves and trees have been used as models for architectural design throughout the history [2]. Biomorphic designs mimicking formal aspects of biological models were extensively used in many architectural styles, for example anthropomorphic floor plans in vernacular building traditions. Natural materials were used to build shelters and houses, and technologies to harvest and process materials were invented and are being continuously developed further until today. Technology is one of the main driving forces for architectural development. The formal transfer of constructions out of ephemeral building materials to durable building materials such as stone is a well-known phenomenon in architecture history, and is a source of information on historical building technology. In Egyptian temples, phytomorphic analogies, forms relating to plants, found in stone columns indicate the use of reed bundles and trunks of palm trees as construction elements, and many other examples refer to symbolic as well as pragmatic use of natures models [2].

The Renaissance genius Leonarda da Vinci, who strategically used nature as a model, is often presented as first biomimeticist. The historical analysis of patents derived by natural models revealed that reinforced concrete, one of the most common modern building materials, was invented by the French gardener Joseph Monier. He experimented with wire mesh to make plant pots more durable after having noticed the sclerenchymatic fibre structure, the strengthening tissue, of decaying parts of opuntia, the prickly pear [1]. This very influential biomimetic invention enabled architects and designers in the last century also to reform nature in organic designs.

Hence in the 1950s, Pier Luigi Nervi, Gio Ponti and Oscar Niemeyer exploited the structural potential of concrete, while others like, for example, Eero Saarinen with his TWA Terminal and Jørn Utzon with the Opera house in Sydney took a more biomorphic direction. The asymptote of design of organic space was Frederic Kessler's so-called Endless house [3].

Findings in natural sciences strongly influence architecture and arts, as can be shown among many other examples by the work of biologist Ernst Haeckel [4] that has been cited and reinterpreted from the turn of the twentieth century until today, for example by the designer Luigi Colani [5]. Representatives of biomimetics in architecture and theorists having researched the overlaps between nature and architecture will be presented in the following.

7.2.1 Analogy and Convergence

The work of Frei Otto and his group in Germany was focused on "natural constructions", a concept encompassing many interesting properties, both in nature and architecture [6]. The concept of "natural constructions" is ambiguous and can include:

- Constructions and structures observable in nature, at all scales and levels of hierarchy, from the nanosurface of the lotus leaf over materials, cells, tissues, organs, populations and ecosystems to the universe itself (if non-living phenomena are included)
- Buildings of animals that have evolved over a long time
- Traditional human building typologies that have developed over a long time
- Architecture showing any natural aspect, for example sustainable design

Frei Otto used an experimental approach that he called "synthetic analogy research" to gain understanding of natural structures and processes like, for example, minimal surfaces and the geometry of soap bubbles and foam. Principles of self-organisation in nature were investigated to be used for form finding. Decades of research and experimental development of methods, structures and materials brought about many very inspiring publications on natural constructions and the development of a novel building technology and typology: modern membrane constructions. The same approach – but with less influence on building technology – was used by the famous Antoni Gaudi and Heinz Isler, who developed a refined technology to build ephemeral shell structures [7, 8]. All three experimented with hanging models to find optimised forms – gravity and self-organisation were used to develop efficient building structures with the least possible amount of material. The search for light constructions culminated in the discovery of the so-called tensegrity structures by Buckminster Fuller and Kenneth Snelson in the 1960s [9]. Tensegrity structures are self-stressing and have remarkable mechanical capacity compared to the amount of material needed. The structures consist of tensioned elements, ropes that create a continuous subsystem and elements under compression, which create a discontinuous subsystem. The construction of vertebrates and the cytoskeleton of cells seem to follow the same construction principle – an interpretation that was found by Donald Ingber, who was inspired by Fuller and Snelson's discoveries [10], and one of the many examples of reverse information flow.

7.2.2 Strategic Search for the Overlaps Between Architecture and Nature

In spite of the obvious importance of the connection between nature and architecture, few attempts to strategically investigate the topic were made. Smaller fields such as zoomorphic and anthropomorphic architecture were explored, the focus lying more or less on the morphological aspect [11, 12].

Paolo Portoghesi compiled the encyclopaedic book "Nature and Architecture", which is not only the most extensive collection of analogies between nature and architecture, but also discusses structural and functional parallels [2].

The first book on "Architekturbionik" was published by the Russian Juri Lebedew in the 1970s and presents a comprehensive collection of then up-to-date architectural developments worldwide, which relied on principles derived from nature, even if their creators did not think of themselves as biomimeticists [13]. Werner Nachtigall published a book on "Bau-Bionik" focusing on analogies in construction, and using structural categories to order natural constructions like fibre reinforced structures, shells, folding structures or membranes [14]. Nachtigall also refers to vernacular architecture as a source of solutions especially concerning climatisation and building physics, and presents current projects that integrate principles like natural ventilation.

Christopher Alexander has described in his life's work "The Nature of Order" an architectural interpretation of life and convergent pattern emergence in nature and architecture, stating the importance of living structures as surrounding for human existence [15]. Alexander is convinced of the necessity of a paradigm shift, from the mechanised processes to widespread overall living processes that involve unfolding wholeness and fundamental differentiation. He defines a living process as any adaptive process that generates living structures, step by step, through structure-preserving transformations. Alexander's approach is not biomimetic, but investigates and applies the topic of life and aliveness in an architectural context.

The "Design and Nature" conferences were launched in 2002, providing a forum for a broad approach exploring nature and its significance to design and architecture [16, 17]. The book "Nature Design" focuses on the relation between nature and the broad field of design, and presents current as well as historic examples. Four essays trace the historical changes in the relation between nature and design from the nineteenth century until today and establish connections to the respective political and economic developments [18].

The author's Ph.D. thesis on "Architekturbionik" [19] is the first strategic search in biomimetics in architecture carried out since Lebedew's groundbreaking work. The use of the so-called criteria of life as an ordering system to categorise current architectural developments is a new approach to investigate the overlaps between nature and architecture. Imposing the life sciences terminology to a technical realm delivers insights that could not be made otherwise and suggests future fields for experimental development. The exploration of the signs of life in architecture has shown that many of our current developments in architecture are based on one or more aspects related to life, for example sensing and reaction, growth, locomotion and others. Morphogenetic and evolutionary design strategies are still key issues in the current architectural discourse and implemented into architectural design by many research groups.

7.3 Strategies: What is Transferred and How is it Done?

7.3.1 What is Transferred?

The transfer of information from one discipline to the other is the most interesting part of the biomimetic process. The transfer of form, the application of morphological characteristics is most common in architecture and design and cannot be excluded from discussion. Furthermore, the symbolic meaning is important, as Sachs puts it ....*that forms inspired by nature become topical when modern society finds itself in crisis, and that the use of organic forms is intended to bring about harmonization and reconciliation with an external world perceived as hostile or uninhabitable* [18]. Even more general than the investigation and transfer of "natural constructions" is the transfer of qualities that can be found in nature. Talking about interesting "natural phenomena" would include for example the play of light falling through foliage, emerging from a complex interplay of different parameters in animate as well as inanimate nature. Nature's phenomena can include surfaces, materials and/or structures, functions, constructions, mechanisms, principles (e.g. self-organisation) or processes (e.g. evolution), delivering models to be analysed, abstracted and applied to architectural solutions on all scales and levels of design (Fig. 7.1).

7.3.2 Methods

The following methods are used in biomimetics in general. Two different approaches to biomimetic information transfer can be discerned according to the direction of information flow respectively starting point of the search.

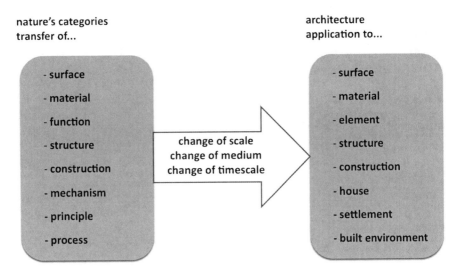

Fig. 7.1 Scheme of nature's categories, information transfer and application, based on [19]

- Biomimetics by induction – bottom-up – solution-based
  - Takes the phenomena found in nature as a starting point, solution-based approach
- Biomimetics by analogy – top-down – problem-based
  - Takes the problem in technology as a starting point, problem-based approach

The terminology "top-down" and "bottom-up" in this respect was introduced by Speck and Harder [20], together with a step-by-step description of activities involved occurring during a biomimetic innovation process. Gebeshuber and Drack criticise that this implies a ranking between nature and technology, and suggest the use of "Biomimetics by Induction", respectively, "Biomimetics by Analogy" instead [21]. Researchers at the Georgia Institute of Technology at the Center for Biologically Inspired Design have described similar processes in detail that they name "solution-based" and "problem-based" approaches [22]. However, all groups describe similar processes, and the experiences made with designing with nature as a model seem to share similarities. The abstraction phase following profound research in the life sciences is considered crucial for the success of transfer. Reduction of complex information and identification of relevant parameters and boundary conditions are necessary for this step. In order to keep options wide at the beginning of information transfer, it is advisable to not categorise too strictly, but to follow the paths of (also very personal) interest using intuition as an important guideline.

Both methods, solution-based as well as problem-based biomimetics, involving both directions of information flow, were used in the design programmes guided by the author, and the implications and outcomes will be discussed in the case studies section (Fig. 7.2).

7 Biomimetics in Architecture[Architekturbionik]

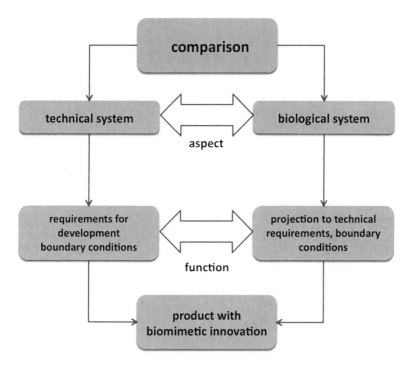

Fig. 7.2 Comparative approach to biomimetic transfer, based on [1]

In the method of Biomimcry, the transfer of functional aspects is the most favoured approach, stemming from the hypothesis that all existing constructions and structures in nature have a functional cause, and that function is the key to the establishment of suitable analogies. In contrast to biomimetics, environmental responsibility and sustainability are directly implemented into the innovation process. Innovation is understood in the notion of a necessity to push industrial developments towards a sustainable future. Biomimicry principles are used as guidelines and evaluation parameters for the innovation process [23].

BioTriz is a general biomimetics-based innovation tool, set up by a group of researchers around Julian Vincent at the Center for Biomimetics in Bath. The system is based on Altshuller's "Theory of Inventive Problem Solving" that was developed in the 1960s based on the analysis and statistical evaluation of technical patents to identify key innovation principles. A TRIZ-based analysis of problem solutions in nature has shown the profound difference of design in nature and design in technology, while technology solves problems largely by manipulating usage of energy, biology uses information and structure [24]. Only recent technologies such as nanotechnology and smart materials work in a more "biological" way. BioTriz is the adaptation of the system to "green" innovation, coming from nature rather than technology [25].

The system of BioTriz and the Biomimicry principles currently gain access to the world of architecture and engineering. One of the largest engineering companies worldwide, Arup, refers to Janine Benyus Biomimicry principles and researchers of Buro Happold that offers multidisciplinary engineering consultancy make use of BioTriz to develop innovative building technology [26, 27].

7.4 Application Fields: Successful Examples

Application fields in architecture are found on all scales, from nanoscale surface definitions over buildings and elements to urban design. The complexity of architecture, its existence on many levels is on the one hand very interesting, as many possibilities for biomimetic innovation exist, but on the other hand it is impossible that all levels are involved in a biomimetic development.

Biomimetic innovation may, for example, only concern technical aspects that are integrated into building technology, or may be limited to the use of a biomimetic surface structure, such as the "Lotusan" paint (a product with self-cleaning properties, developed by the STO company in 1999 based on the "Lotus-effect" patent of Prof. Wilhelm Barthlott).

For this reason, the visibility of a biomimetic approach in the final design is not guaranteed, and not always aesthetic quality is involved. On the other hand, "bionic architecture" starts to be introduced in the public as a new kind of what used to be called "organic architecture", and "bionic" is increasingly used for advertising purposes. Holistic visions of a biomimetic architecture range between naïve romanticism and futuristic experimental design. The design proposals for a "Vertical Garden City", the so-called Bionic Tower of two Spanish architects, were promoted as the first existing model of a "bio-ecological urban structure", in spite of numerous less-known serious attempts to sustainable design in high-density structures [28]. Based on flexibility principles and biological structures, it is said to be adapted to the different economic, environmental and social conditions of the cities where it is built. But the definition of a "biomimetic architecture" as a new style seems as impossible as that of a "bionic car", a concept car developed using the formal analogy of the boxfish and biomimetic optimisation strategy. What can be stated in this example is valid for architecture as well as other disciplines: a biomimetic approach to novel solutions concerning a limited number of aspects of the design is used. For the different architectural scales, selected recent architectural examples are presented in the following, with regard to the qualities that have successfully been transferred.

7.4.1 Emergence and Differentiation: Morphogenesis

Emergence and differentiation are in the focus of the "Emergent Technologies and Design" masters programme at the Architectural Association in London. Michael Weinstock, Michael Hensel and Achim Menges' team elaborate refined structures in close collaboration with experts in biomimetics, mathematics, structural engineering and material science, investigating novel technologies. Material systems are developed, which integrate form, material and structure according to the structures found in nature. The projects that were developed by the students within this frame explore, for example, dynamic relations and behaviour of occupation patterns, environmental modulations and material systems, in natural as well as computational environments. Morphogenesis is investigated as a new design strategy, based on dynamic adaptation processes [29–31].

7.4.2 Interactivity

Kas Oosterhuis and his Hyperbody Research Group at the TU Delft explore what they call the "building body" [32].

Protoarchitectural projects experiment with reactivity to environmental influence and control. "Hyperbodies are pro-active building bodies acting in a changing environment. The HRG [Hyperbody Research Group] introduces interactivity not only in the process of collaborative design, but also during the use and maintenance of buildings" [33]. The biomimetic fluidic muscle finds application in architecture as easy to implement pneumatic actuation system for kinetic systems. One example is Oosterhuis' exhibition project "Muscle", a pressurised soft volume wrapped in a mesh of tensile Festo muscles, which can change their own length [34]. The muscles generate an apparent uncontrolled overall behaviour by moving according to programmed control, and in interaction with the visitors [35, 36]. Interactivity on a symbolic level is exploited by numerous architectural projects around the "media facade", which may be biomimetic due to reactivity and/or control.

7.4.3 Dynamic Shape

Shape change in architecture is the result of a morphogenesis, which is not static any longer but a constant dynamic process of interaction between the building and the environment. One of the first built structures that actually changes internal space is Transformer, a temporary, shape-shifting structure designed by Rem Koolhaas in Seoul, South Korea. The pavilion consists of a steel structure wrapped in a translucent elastic membrane. The 20-m high structure has to be picked up by cranes to be rotated [37]. An unbuilt but widely published project is the "Rotating Skyscraper" by David Fisher, to be realised in Dubai. The idea of changing orientation by rotation is not new, but the dimension of the project exceeds the scale of rotating architectures that were built until today (e.g. the single family house Heliotrop in Freiburg, Germany). The Rotating Skyscraper consists of a central core, with the single floors cantilevering and rotating independently, thus changing the orientation of the floor plans and the overall form and appearance of the building [38]. Experiments for adaptive structures are carried out at the Institute for Lightweight Structures and Conceptual Design in Stuttgart. Patrick Teuffel investigated the adaptation of structures by changing the stiffness or lengths of individual elements to manipulate the flow of forces. He developed a design concept for adaptive systems, the so-called load path management, a system that dynamically adapts shape to load, thus replacing mass through energy in using a "virtual" stiffness due to actuation instead of physical stiffness [39]. These projects represent the validity of the biological paradigm in current architectural development, without directly using a biomimetic approach.

7.4.4 Intelligence

Control is also an important issue in "intelligent buildings" that are intended to provide an environment that adapts itself to the needs of the inhabitant in many ways. One way to implement smartness in buildings is the use of smart elements or materials that can react to environmental influence. Research and development in intelligent building technology is focused on integration of sensors in architecture, development of heating, cooling and ventilation systems, and on the overall control of this environment, in interaction with the user [40]. Biological models for the behaviour of integrated control systems are investigated by many research groups, also at the Vienna University of Technology, for example by the Institute of Computer Technology [41].

7.4.5 Energy Efficiency

Energy efficiency is one of the most important aspects linking biomimetic visions with living nature. As building facades represent the interface between internal space and environment, facade technology is a focus of research in energy efficiency of architecture. Biomimetic approaches are manifold and range from diverse adaptations of cacti to hot climate and intense solar radiation [42, 43] to ivy as model for a solar energy harvesting system, which can be added to an existing facade [42, 44]. Salmaan Craig developed energy-efficient building surfaces using BioTriz, and in this way working with a variety of models from nature delivering functional solutions [27]. Lidia Badarnah is part of the Facade Research Group at the TU Delft that investigates the use of biomimicry as innovation tool mainly for building envelopes [45]. Dirk Henning Braun compiled an analysis of different role models from nature for a visionary adaptive permeable skin structure [46].

Apart from facade concepts, ventilation is another key issue. Natural models like the termite mound and other passive ventilation systems in nature inspire innovative building technologies, even if the phenomena may not yet be fully understood [47].

7.4.6 Material/Structure/Surface

The differentiation between material, structure and surface is no longer valid when working with nature as a model, which is important also for biomimetic approaches to energy efficiency of facades. Research and development takes place on more than one scale; so the topic of energy efficiency is strongly connected to the influence of nanotechnology in architecture. Nanotechnology is the discipline of researching and manipulating materials on the scale of molecules and atoms, providing entirely new possibilities for the development of materials with desired properties. The use of smart materials that can react to changing environmental conditions has already become common in building industry. Self-cleaning and easy-to-clean surfaces are applied in the glass industry and as coatings of construction materials and products. Other functions of already available nanotechnological surface coatings include anti-reflectivity, switchable transparency and darkening in photochromic glass, anti-fingerprint, fire protection, antibacterial, scratch proofness, air cleaning and microcapsules for fragrances [48]. Self-healing, self-repair and autonomous energy supply are functions that already require nano-stucturing beyond surface coating. Self-organised healing and repair of elements are especially interesting in cases where local failure would lead to a total system breakdown, as in airplanes, space technology or in pneumatic structures, which rely on air pressure to maintain structural integrity. Self-repair according to natural self-repair mechanisms in plants is explored by the Plant Biomechanics Group in Freiburg and applied to polymer materials and membranes for pneumatic constructions [49]. Self-organised crack polymerisation is experimented with for use in space applications and in concrete. Hollow glass fibres filled or microspheres with polymer matrix are integrated into composite materials. The breakage of the containers and the following release of the sealing matrix fulfil the tasks of detection and solution at the same time [50]. Biological models for effective reversible and high adhesion like the gecko foot and seashells are experimented with [51] also for expected implementation in building industry.

7.4.7 Integration

The topic of integration has a diversity of notions in the context of biomimetics in architecture. Integration of material, structure and surface is not a new idea, but new micro- and nanosystems production technologies boost developments in this field. The goal of a research project at the Massachusetts Institute of Technology in the USA that explores hierarchical combination of Carbon nanotubes is to generate a material architecture that governs a combination of structural, optical and fluidic behaviour. "Construction *in vivo*" is a novel approach to designing, fabricating and maintaining building skins by controlling the mechanical and physical properties of spatial structures inherent in their microstructure. This material can be differentiated and adapted according to local and global context [42, 52]. On a macro-scale, fibre composite materials lend themselves for the production of hierarchical material and the integration of functional elements. Research at the ITV Denkendorf in Germany explores the production of differentiated fibre composite structures with pultrusion technology, following principles derived from plants [53]. The technology of pultrusion (a synthesis of the terms pull and extrusion) that produces composite fibres with continuous cross section is combined with a braiding technology to create three-dimensional structures.

Integration of nature into architecture seems to increase with the density urban structures are populated – developments towards integrative concepts come from countries with very high population density: Ken Yeang's concepts of green skyscrapers, which intend to enhance quality of life and integrate architecture into a cycle of resource and energy flow, were designed for urban areas in southeast Asia. The architecture firm MVRVD's world expo 2000 project of a stacked landscape came from the Netherlands, one of the most densely populated countries in Europe. Building with nature is another integrative concept, which is used by the so-called Baubotanik group in Stuttgart, which makes use of living trees for architectural structures like bridges and pavilions, together with a biotechnological approach concerning the connections and integration of technical elements [54].

7.5 Case Studies

The author has more than 8 years of experience in biomimetics in architecture, comprising guidance of students projects at the architecture faculty of the Vienna University and Technology, design programmes, workshops and academic studies [19, 55, 56]. The outline of the programmes and a selection of resulting projects will be presented and analysed in the following.

7.5.1 Biomimetics Design Exercise

Biomimetics was introduced at the "Department for Design and Building Construction" at the Vienna University of Technology as part of a special education in membrane constructions. Within this frame, a combination of lectures and design exercises were carried out, using a thorough inductive approach, based on the personal fascination for a natural phenomenon [55]. Approximately 60 students per year attended the course, and about 25% of the design exercises were successful in delivering innovative architectural concepts (even if far from being realised). Challenges for the students ranged from selection of an appropriate model to finding of scientific information, interdisciplinary scientific research, understanding and abstraction of a principle to the identification of an application field and scenario. In many cases, scientific information about the selected phenomenon was not

7 Biomimetics in Architecture[Architekturbionik]

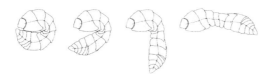

Fig. 7.3 Geometrical analysis of pill bug [57]

Fig. 7.4 Model of pill bug shell, open and coiled up state [57]

accessible. In particular, when spatial issues in tangible scale were of interest, the students themselves carried out research. In some cases, this basic research exhausted the timeframe of the programme, but delivered surprising insights for further studies. The examination of the geometrical characteristics of the pill bug, for example, delivered insights about the functional morphology of the segmented outer shell. The geometry of the segments and the position of the pivot points lead to the almost spherical form of the coiled up animal [19, 57]. This is a very interesting system considering the efforts of mathematicians and designers to define patterns for the tiling of spheres and curved shapes in general (Figs. 7.3 and 7.4).

Interest in spatial development or dynamics was characteristic for many natural phenomena that were selected by the students. A successful translation of crystal growth was carried out by Kieser/Oberndorfinger, who developed a pneumatic structure that could change its overall shape [19, 58]. The system could be applied to create a flexible space (Figs. 7.5–7.7).

The research of adaptive systems and the transfer to architectural applications were issues that many students were interested in, obviously also inspired by the current architectural discourse. Smart, reactive and interactive elements were designed inspired by biological mechanisms investigated in a variety of plants and animals. The inspiration for the experimental design project "aero dimm", a pneumatic facade with colour change option, came from the colour change of cephalopod skin: Between two layers of a facade, elastic membranes change volume by pneumatic pressure, creating dark parts on the surface. The ventilation tubes can be integrated into the inside membrane of the system. "Aero dimm" could be used for facade darkening as well as colour change (Fig. 7.8) [19, 59].

Reasons for failure in achieving a biomimetic transfer were manifold – simple disinterest in nature, limited knowledge, selection of unexplored phenomena,

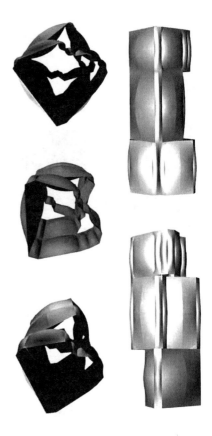

Fig. 7.5 Pneumatic structure, inspired by crystal growth processes [58]

superficial examination, inability to identify abstract principles, misinterpretation of the phenomenon and scaling problems to name a few. Successful transfers were characterised by a focused research phase and precise abstraction of a principle to be applied in architecture.

7.5.2 Biomimetics Design Programmes, Workshops and Studies

The design programmes were carried out at the Department for Design and Building Construction at the Vienna University of Technology. The introduction of biomimetics in larger design programmes required the introduction of a problem-based method, as given architectural tasks had to be dealt with. The usual timeframe for a design programme is one semester. In all programmes, the approach taken in was a combination of problem- and solution-based approach that included the following steps.

First, a broad search for interesting phenomena was used, which identified groups of organisms and aspects that were assumed to be in some way interesting for the given task. A research phase delivered scientific information on these phenomena.

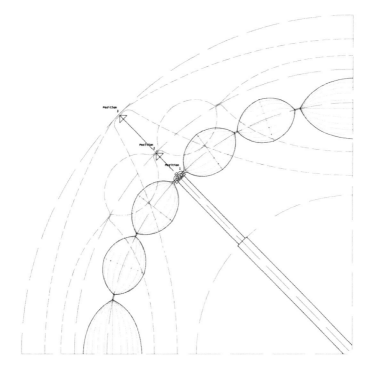

Fig. 7.6 Construction detail [58]

The many options were then gradually narrowed down using evaluation systems that were based on the boundary conditions of the architectural task. The selected examples were explored in depth, effective principles identified and applied to a specific architectural assignment.

Two projects aimed at the development of architectural solutions for outer space: "Transformation structure/space" (2004) and the study "Lunar Exploration Architecture" (2006), a study for the European Space Agency [60,61]. Biomimetics was used to find innovative designs for deployable structures for a lunar base.

The workshop "Bioinspired facade design" was carried out in 2007 in collaboration with the London-based architecture firm Horden Cherry Lee Architects and a British surfboard company, Beach Beat surfboards. A team of architects and engineers accompanied the programme from the Vienna University of Technology. In contrast to other case studies, the task was clearly defined: the scope was to design a specified part of facade cladding for a high rise building, with a specific material technology that was taken from the sports industry, where high-performance materials are commonly used and developed. Based on investigations on repetitive structures in nature, for example locomotion rhythms of earthworm [62], bark of trees, fractal structure of human lungs, etc., and on fibre composites technology, four different design projects were developed until prototype stage (Figs. 7.9–7.11).

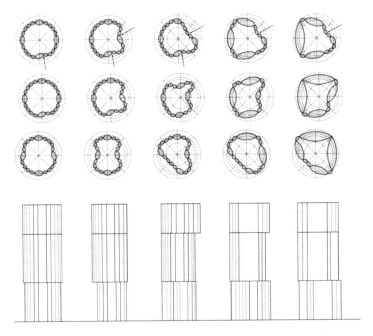

Fig. 7.7 Possible variations of the pneumatic space, sections and elevations [58]

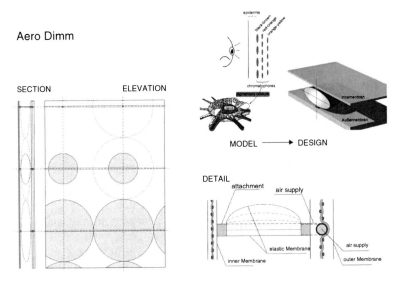

Fig. 7.8 "Aero Dimm" – inspired by cephalopods, pneumatic facade darkening [59]

7 Biomimetics in Architecture[Architekturbionik] 143

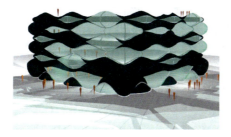

Fig. 7.9 "Worm" project, bioinspired facade design workshop, visualisation [62]

Fig. 7.10 Prototype model of facade element, fibre composite material [62]

Fig. 7.11 Horizontal section of "Worm" facade [62]

A comparison of all design exercises and programmes mentioned is presented in the following table. For the comparison, the starting points, the transfer and the endpoints of the four different studies are considered. The kind of programme relates to the design task, intensity of research, availability of experts and duration of work (Fig. 7.12) [56, 63].

The implementation phase often brought about new challenges and problems, which was again handled in a biomimetic approach by the students, without being instructed to do so. In this way, added biomimetic ideas were implemented in the designs, which is being referred to as "added biomimetic approaches" in the comparison table. This phenomenon was also observed in the studies of the "Center for Biologically Inspired Design" at the Georgia Institute of Technology in the USA and integrated into the method of the so-called compound analogical design [64]. Consultancy of experts in the life sciences and/or biomimetic transfer was experienced as very important and always stimulating. Growing abstraction of a natural phenomenon increases applicability and transfer, but can remain superficial when research was not profound enough.

	Biomimetic design exercise	Transformation architecture	Lunar exploration architecture	Bioinspired facade design
Kind of program	course in biomimetics	design program	study	design program
Starting points	topics out of nature according to individual interest	big topics like folding mechanisms, functional anatomy	different particular topics out of nature	grouped but particular topics of out of nature
Design task	architectural application	space application	lunar exploration	facade cladding
Task defined	after research	parallel to research, scenario definition	parallel to research, scenario definition	task presented after first research steps
Intensity of research	from superficial to intense literature research to own experimental research	intense literature research	intense literature research, own experimental research	literature research
Transfer of	function, construction, mechanism, material, locomotion, surface structure, spread, production process	mechanism, functional anatomy, function, locomotion	folding and deployment mechanism	repetitive form, geometry, rhythm of locomotion, function
Added biomimetic approaches	yes/no	yes	no	yes sustainability
Experts involved	no / seldom	biomimetics and space experts	biomimetics, mechanical engineering and space experts	no
Timeframe	2 months	1 semester	4 months	1 semester
Number of successful projects	133	4	1	4
Success rate	25%	50%	100%	100%
Difficulties	finding of topic, abstraction	development of scenario together with application field	development of scenario together with application field	abstraction

Fig. 7.12 Comparison of programmes at TU Vienna regarding important aspects of transfer and conditions

The development of an application scenario (e.g. in the space projects) was experienced as an obstacle, and took much time and effort. On the other hand, can a specified task result in oversimplified transfers?

As a result, the strategy to only roughly define an application field delivers the most innovative results, but cannot be applied to particular tasks. Respective time has to be allocated to the development of the application scenario.

Gradual differentiation of the design with respect to starting points and application may deliver an applicable method for biomimetic architectural design.

7.6 Future Fields, Aims and Conclusion

7.6.1 Aims

Even if the biomimetic approach is not carried out thoroughly from analysis in the life sciences to technical application, the use of models taken from nature meanwhile seems to be a commonly used approach in current architectural design.

Within the last decades, the use of biology's paradigm in architecture has increased. *... it is now becoming evident that a new strand of biomorphism is emerging where the meaning derives not from specific representation but from a more general allusion to biological processes... [Biomorphism] signifies the commitment to the natural environment... is apt to be considered more environmentally responsible (or responsive)...* [11]. There seems to be a general belief that the use of biological models will help to develop a more sustainable and ecologically responsible future, which leads to a better quality of life in built environment.

It is beyond controversy that the discussion of nature and natural technologies delivers increased knowledge and consciousness about ecological interconnections, but biomimetics as a sole innovation tool can also deliver unsustainable products and is not a panacea for all global problems [65]. The efforts of biomimetic architectural development should be targeted to sustainability and energy and resource efficiency, but application of a biomimetic innovation method does not guarantee an ecological solution. Even if the investigation of nature's principles imply increased consciousness of natural processes and complexity, the aim to design environmentally responsible has to exist independently from the innovation method. Current developments in biomimetics in architecture and design also include networking on an international level.

Scientific and commercial interests in biomimetics have led to the founding of several organisations, such as the Bionc Engineering Network BEN in Germany [66], BIOKON international based in Germany [67] with a group focusing on architecture and design, BIONIS in the UK [68] with a focus on engineering and the Biomimicry group in the USA [69].

7.6.2 Considerations About Future Developments

The one to one modelling, the creation of prototypes of new architectural solutions is still important – the virtual modelling capacities and qualities do not yet reach a stage where sole digital development would be sufficient to grasp the future qualities of a new architectural solution. Until now, limited factors of buildings or projects are investigated digitally, for example the influence of the sun, acoustics, internal stresses due to loads and environmental forces. Moreover, consultants work on these factors separately, considering different scales and different computer modelling tools.

Digital models, on the other hand, allow the introduction of design algorithms from nature: optimisation processes, growth processes, evolutionary processes for example can be applied. The processes are relocated from the real world to a digital model space. New production processes influence and determine the creation of architecture. In the same way design and production will be modified by the biological paradigm, when biomimetic principles are applied.

Marcos Novak, investigating actual, virtual and mutant intelligent environments states: *One of the fundamental scientific insights of this century has been the realization that simulation can function as a kind of reverse empiricism, the empiricism of the possible. Learning from the disciplines that attend to emergence and morphogenesis, architects must create generative models for possible architectures. Architects aspiring to place their constructs within the nonspace of cyberspace will have to learn to think in terms of genetic engines of artificial life. Some of the products of these engines will only be tenable in cyberspace, but many others may prove to be valid contributions to the physical world* [70].

The future of biomimetics in architecture is the creation of visions with the strength to establish innovation for the improvement of the quality of our built environment.

References

1. W. Nachtigall, *Bionik, Grundlagen und Beispiele für Ingenieure und Naturwissenschaftler* (Springer, New York, 2002)
2. P. Portoghesi, *Nature and Architecture* (Skira editore, Milan, 2000)
3. Historisches Museum der Stadt Wien, D. Bogner (ed.): Friedrich Kiesler 1890–1965: inside the endless house (Böhlau Wien Köln Weimar, 1997)
4. E. Haeckel, *Kunstformen der Natur, Neudruck der Farbtafeln aus der Erstausgabe "Kunstformen der Natur" Leipzig und Wien, Bibliograpisches Institut, 1904* (Prestel-Verlag, München, 1998)
5. P. Gruber, The signs of life in architecture. IOP Bioinspir Biomim **3** (2008)
6. F. Otto et al., *Natürliche Konstruktionen, Formen und Strukturen in Natur und Technik und Prozesse ihrer Entstehung* (Deutsche Verlags-Anstalt GmbH, Stuttgart, 1985)
7. R. Zerbst, *Antoni Gaudi* (Benedikt Taschen Verlag, Köln, 1987)
8. H. Isler et al., *Heinz Isler Schalen: Katalog zur Ausstellung* (Paris, 1986)
9. R.B. Fuller, J. Krausse, C. Lichtenstein, *(Hg.): Your private sky Richard Buckminster Fuller, Design als Kunst einer Wissenschaft* (Verlag Lars Müller und Museum für Gestaltung, Zürich, 1999)
10. D. Ingber, Die Architekturen des Lebens, in ARCH+ Verlag GmbH Nikolaus Kuhnert, Sabine Kraft, in *ARCH+ Zeitschrift für Architektur und Städtebau Nr.159 160, Formfindungen von biomorph bis technoform*, vol. 5, ed. by G. Uhlig (ARCH+ Verlag GmbH, Aachen, 2002), pp. 52–55
11. H. Aldersey-Williams, *Zoomorphic, New Animal Architecture* (Laurence King Publishing, London, 2003)
12. G. Feuerstein, *Biomorphic Architecture, Menschen- und Tiergestalten in der Architektur* (Edition Axel Menges, Stuttgart, 2002)
13. J.S. Lebedew, *Architektur und Bionik* (Verlag MIR, VEB Verlag für Bauwesen, Moscow, 1983)
14. W. Nachtigall, *Bau-Bionik, Natur, Analogien, Technik* (Springer, New York, 2005)
15. C. Alexander, *The Nature of Order, The Process of Creating Life* (The Center for Environmental Structure, Berkeley, 2002)
16. C.A. Brebbia, M.W. Collins (eds.), *Design and Nature II* (Wessex Institute of Technology, Wessex, 2004)
17. C.A. Brebbia (ed.), *Design and Nature III, Comparing Design in Nature with Science and Engineering* (Wessex Institute of Technology, Wessex, 2006)
18. Museum für Gestaltung Zürich, S. Angeli (ed.), *Nature Design, From Inspiration to Innovation* (Lars Müller Publishers, Baden, 2007)
19. P. Gruber, *Biomimetics in Architecture [Architekturbionik] – The Architecture of Life and Buildings*, Doctoral thesis, Vienna University of Technology, 2008

20. T. Speck, D. Harder, M. Milwich, O. Speck, T. Stegmaier, Die Natur als Innovationsquelle, in *Technische Textilien*, ed. by P. Knecht (Deutscher Fachverlag, Frankfurt, 2006) pp. 83–101
21. I.C. Gebeshuber, M. Drack, An attempt to reveal synergies between biology and engineering mechanics, invited article, Proc. IMechE Part C: J. Mech. Eng. Sci. **222**, 1281–1287 (2008)
22. S. Vattam, M. Helms, A. Goel, Biologically Inspired Innovation in Engineering Design: A Cognitive Study. Technical Report 20 S.S. Vattam, M.E. Helms, A.K. Goel, GIT-GVU-07–07, Graphics, Visualization and Usability Center, Georgia Institute of Technology, 2007
23. www.biomimicry.net/. Accessed Oct 2009
24. J. Vincent, O. Bogatyreva, N. Bogatyrev, A. Bowyer, A. Pahl, Biomimetics – its practice and theory, J. R. Soc. Interface **3**(9), 471–482 2006
25. www.biotriz.com/. Accessed Oct 2009
26. P. Head, The Institution of Civil Eningeers Brunel Lecture Series: Entering the ecological age: the engineer's role, Arup, 2008
27. S. Craig, D. Harrison, A. Cripps, D. Knott, BioTRIZ suggests radiative cooling of buildings can be done passively by changing the structure of roof insulation to let longwave infrared pass, J. Bionic Eng. **5**, 55–66 (2008)
28. www.cerveraandpioz.com. Accessed Oct 2009
29. International House, *London: AD Architectural Design, vol. 74, no 3, Emergence Morphogenetic Design Strategies* (Wiley, Chichester, 2004)
30. International House, *London: AD Architectural Design, vol 76, no 180, Techniques and Technologies in Morphogenetic Design* (Wiley, Chichester, 2005)
31. M. Hensel, A. Menges (eds.), *Form follows performance - zur Wechselwirkung von Material, Struktur und Umwelt*, Arch+ Nr.188 (Arch+ Verlag Aachen, 2008)
32. K. Oosterhuis, *Hyperbodies, Towards an E-motive Architecture* (Birkhäuser, Basel, 2003)
33. www.bk.tudelft.nl [10/2009]
34. www.festo.com/INetDomino/coorp_sites/de/e2e6ed26f15af734c12572d20065650c.htm. Accessed Jan 2010
35. www.oosterhuis.nl/quickstart/index.php?id=347. Accessed Oct 2009
36. H.M. Mulder; L.A.G. Wagemans, An Adaptive Structure Controlled by Swarm Behaviour, IASS Symposium, Montpellier, 2004
37. www.oma.nl. Accessed Oct 2009
38. www.dynamicarchitecture.net. Accessed Oct 2009
39. P. Teuffel, *Entwerfen Adaptiver Strukturen, Lastpfadmanagement zur Optimierung Tragender Leichtbaustrukturen*, Dissertation, Institut für Leichtbau Entwerfen und Konstruieren Universität Stuttgart, 2004
40. D. Clements-Croome (ed.), *Intelligent Buildings* (ICE Publishing, London, 2004)
41. D. Dietrich, G. Fodor, G. Zucker, D. Bruckner (eds.), *Simulating the Mind, A Technical Neuropsychoanalytical Approach* (Springer Verlag Wien, 2009)
42. T. Klooster (ed.), *Smart Surfaces, Intelligente Oberflächen und ihre Anwendung in Architektur und Design* (Birkhäuser, Basel, 2009)
43. www.case.rpi.edu. Accessed Oct 2009
44. www.s-m-i-t.com. Accessed Oct 2009
45. L. Badarnah, U. Knaack, Organizational features in leaves for application in shading systems for building envelopes, in *Proceedings of the Fourth Design & Nature Conference: Comparing Design and Nature with Science and Engineering*, Southampton, 2008, ed. by C.A. Brebbia, pp. 87–96
46. D.H. Braun, *Bionisch Inspirierte Gebäudehüllen*, Dissertation, Institut für Baukonstruktion Lehrstuhl 2, Stuttgart, 2008
47. J.S. Turner, R.C. Soar, Beyond biomimicry: What termites can tell us about realizing the living building, in *First International Conference on Industrialized, Intelligent Construction (I3CON)*, Loughborough University, 2008
48. I.C. Gebeshuber, M. Aumayr, O. Hekele, R. Sommer, C.G. Goesselsberger, C. Gruenberger, P. Gruber, E. Borowan, A. Rosic, F. Aumayr, Chapter IX: Bacilli, green algae, diatoms and red blood cells – how nanobiotechnological research inspires architecture, in

Bio-Inspired Nanomaterials and Nanotechnology, ed. by Y. Zhou (Nova Science Publishers, Hauppauge, 2010)
49. T. Speck et al., Self-repairing membranes for pneumatic structures, in *Proceedings of the Fifth Plant Biomechanics Conference*, 2006
50. R. Trask et al., *Enabling Self-Healing Capabilities – A Small Step to Bio-Mimetic Materials* (ESA, 2005)
51. B. Bushan, Biomimetics: lessons from nature – an overview, Phil. Trans. R. Soc. A 28 April 2009 **367**(1893), 1445–1486 (The Royal Society London 2009)
52. www.nanobliss.com. Accessed Oct 2009
53. M. Milwich, T. Speck, O. Speck, T. Stegmaier, H. Planck, Biomimetics and technical textiles: solving engineering problems with the help of nature's wisdom, Am. J. Bot. **93**, 1455–1465, 2006
54. www.baubotanik.de. Accessed Oct 2009
55. P. Gruber, Transformation architecture, in *Fortschritt Berichte VDI, First International Industrial Conference Bionik 2004, Hannover Reihe 15*, ed. by R. Bannasch, I. Boblan (VDI Verlag GmbH, Düsseldorf, 2004), pp. 13–21
56. P. Gruber, Transfer of nature to architecture – analysis of case studies, in *Proceedings of the Biological Approaches for Engineering*, Southampton, 2008
57. K. Fuchs, *Pillbug Shell, Design proposal for Biomimetics Design Exercise* (TU, Vienna, 2005)
58. S. Kieser, J. Oberndorfinger, *Pneumatic Crystal, Design Proposal for Biomimetics Design Exercise* (TU, Vienna, 2006)
59. S. Pfaffstaller, *Aero Dimm, Design Proposal for Biomimetics Design Exercise* (TU, Vienna, 2004)
60. P. Gruber, B. Imhof, Transformation: structure/space studies in bionics and space design, Acta Astronautica **60**(4–7), 561–570 (2006)
61. P. Gruber, B. Imhof, S. Häuplik, K. Özdemir, R. Waclaviceka, M.A. Perino, Deployable structures for a human lunar base, Acta Astronautica **61**(1–6), 484–495 (2007)
62. A.E. Crebelli, S. Diwischek, A. Dolapcioglu, *Worm Project, Surfcornwall Design Program* (TU, Vienna, 2007)
63. P. Gruber, Bioinspired architectural design, in *Bionik: Patente aus der Natur, Vierter Bionik Kongress Hochschule Bremen*, ed. by A. Kesel, D. Zehren (2009)
64. S. Vattam, M. Helms, A. Goel, J. Yen, M. Weissburg, Learning about and through biologically inspired design, in *Proceedings of the Second Design Creativity Workshop*, Atlanta, June 22, 2008, http://home.cc.gatech.edu/dil/uploads/DCC-Creativity.pdf
65. I.C. Gebeshuber, P. Gruber, M. Drack, A gaze into the crystal ball – biomimetics in the year 2059, Proc. IMechE Part C: J. Mech. Eng. Sci 223, 2899–2918 (2009) (Special Issue Paper)
66. www.b-e-n.eu. Accessed Jan 2010
67. www.biokon-international.com. Accessed Jan 2010
68. www.extra.rdg.ac.uk/eng/BIONIS. Accessed Jan 2010
69. www.biomimicry.net. Accessed Jan 2010
70. M. Novak, Transmitting architecture: the transphysical city, http://www.ctheory.net/articles.aspx?id=76. Accessed Nov 2007

Chapter 8
Biomorphism in Architecture: Speculations on Growth and Form

Dörte Kuhlmann

Abstract Many of the design methods applied by the current architectural avant-garde can be traced back to one of the oldest and most influential ideas in architectural history: the concept of organicism in its various guises. The basic idea of organicism, to take nature as model, is one of the most oldest and most fundamental aesthetic concepts in western art and architecture theory. Since the Renaissance, it has shown an uninterrupted continuity, influencing architecture on both the conceptual and the metaphorical levels. Not only classical but also modern architects attempted to imitate natural forms or processes in design. While the influence of classical philosophers waned during and after the Enlightenment, the appeals to the authority of nature only intensified. Thus, the study of organicism concerns a very basic question in the history of architectural theory as well as in the current discourse.

8.1 Introduction

The form, then of any portion of matter, whether it be living or dead, and the changes of form which are appearent in its movements and in its growth, may in all cases alike be described as due to the action of force. In short, the form of an object is a 'diagram of forces', in this sense, at least, that from it we can judge of or deduce the forces that are acting or have acted upon it [1]. The basic idea of biomorphism, bionics, and organicism in architecture, to take nature as model, is one of the most oldest and most fundamental aesthetic concepts in western art and architecture theory. Since the Renaissance, it has shown an uninterrupted continuity, influencing architecture on both the conceptual and the metaphorical level. As late

D. Kuhlmann (✉)
Vienna University of Technology, Karlsplatz 13, 1040 Vienna, Austria
e-mail: d.kuhlmann@tuwien.ac.at

as 1747, French philosopher and aesthetic Charles Batteux still reduced the fine arts to a single principle, that of *ars imitatur naturam*.

The Aristotelian theory of organic form, as defined by Friedrich Schlegel and Samuel Taylor Coleridge (and restated by some of the most famous twentieth century architects, Frank Lloyd Wright, Hugo Häring, Le Corbusier, and other architects following functionalist ideas), implies that the *unfolding* of the innermost essence of a being is the source of value, and that any outside influences could only be harmful. Aristotle does not hesitate to reason, for example, that "if the movement of the soul is not of its essence, movement of the soul must be contrary to its nature" [2]. This line of organicism sponsored functionalism and led naturally to an interest in "authorless" planning, biomorphic strategies, vernacular traditions, and user planning, all explored by many architects over the past decades. It also underlies more recent avantgarde concepts such as the "death of the author," which redefine the role of the architect and provide a philosophical foundation for contemporary design methods that use computer-generated forms and design algorithms while referring to natural sciences.

The obvious difficulty in designing from the inside out is to determine what the essential nature of a building is. While Peter Eisenman's early houses or Greg Lynn's projects have occasionally been understood as proposing the building as an end in itself, most architects choose to see a house as a reflection, expression, or extension of its inhabitant. In the *Nicomachean Ethics*, Aristotle maintained that "handicraft is he that made it," implying that the limits of an entity are not the limits of its immediate body but that houses and machines, for example, are parts of living beings, extensions of man [3]. Following this suggestion, a house should be read neither as "dead matter" nor as a natural entity but as a natural secretion of a person in the same sense that a snail secretes its shell.

Not only classical but also modern architects attempted to imitate natural forms or processes in design. Sir James Hall proposed already around 1793 a neogothic willow cathedral constructed of living trees, which was rebuilt for the IGA in Rostock 2003 (Fig. 8.1). While the influence of classical philosophers waned during and after the Enlightenment, the appeals to the authority of nature only intensified. Thus, the study of organicism and bionics in architecture concerns a very basic question in the history of architectural theory as well as in the current discourse.

8.2 The Essence of Nature

Despite a long line of various traditions of *ars imitatur naturam*, nature has not always been considered as a purely positive realm. We do not have to go far in history to find vivid attacks of nature such as postulated by Leopardi:

> *Now despise yourself,*
> *Nature, you brute force*
> *who furtively ordain universal doom,*
> *and the infinite futility of all existence* [4].

Fig. 8.1 James Hall, Willow Cathedral, credit: RIBA Library Drawings Collection

With these terse and chilling words, Giacomo Leopardi ended his poem "To Himself" in early 1833 [4]. Although the poem hit a new lugubrious low in western letters, Leopardi was not proposing anything novel. In his resentment of nature, he was rather perpetuating a tradition which goes back at least to the Orphic doctrine of *soma sema*, according to which the body (and by extension all nature) is a cruel and alien prison from which the human soul struggles to break free to return to its proper spiritual home.

Yet, the overwhelmingly dominant tradition of Western thought is predicated on the confident identification of the natural with the good. Seventy-five years before Leopardi's poem, Denis Diderot stated categorically that "water, air, earth and fire, everything in nature is good. Even the gale, at the end of autumn, which rocks the forests, beats the trees together, and snaps and separates the dead branches; even the tempest, which lashes the waters of the sea and purifies them; even the volcano, casting a flood of blazing lava from its gaping side, and throwing high into the air the cleansing vapour." Rejecting anthropocentric ethics, Diderot defined the good as that which comes from nature as opposed to anything devised by man who has been "perverted by the wretched customs of society" [5].

The inherent goodness of nature has been accepted by most classical thinkers who have sought to ground both ethics and aesthetics on the example of nature. Aristotle's definition of art as the imitation of nature provided the unquestionable premise for two millennia of classicism.

Looking at the most recent developments in architecture, there are certainly different ways the Aristotelian principle of *ars imitatur naturam* has been applied, both in architectural theory and in design. Besides the Platonic, imitation of the forms of natural beings, many architects picked up issues of anthropomorphic proportion, the doctrine of form and function being interdependent, the implications of ecological thinking for architectural theory, the influence of natural sciences, and, finally, the very concept of organic unity.

8.3 Nature as a Source for Form

Like the current avantgarde, architects of the past turned to contemporary debates in philosophy and natural sciences to develop new design strategies. In architecture, the notions of organicism led to the rather Platonic imitation of natural forms (Mimetics). Callimachus, the alleged inventor of the Corinthian order, is credited for designing a bronze chimney in the form of a palm tree in the Erechtheion. The application of plant and animal shapes to ornaments constitutes perhaps the most obvious case of architecture imitating nature, but the principle was not always limited to small details. Claude-Nicholas Ledoux designed a phallic-shaped brothel (1792) and Lequeu a dairy in the form of a cow (1790). Later, Rudolf Steiner married Callimachus and Ledoux in his design for the heating plant in Dornach (1914); Herb Greene built his vacation house in a shape evocative of a buffalo or a "prairie chicken," (1960) and Imre Makovecz gave the dormers in his buildings eyelashes (1982).

Even though modern functionalist architects rejected naturalist ornaments, they still found in nature "the eternal example for every human creation," to quote Walter Gropius. Not only for Frank Lloyd Wright but also for his arch-enemy Le Corbusier, Mother Nature was the "great and eternal teacher" [6]. Yet the lessons they learned from their eternal *alma mater* were radically different. Even poststructurally oriented theorists, such as Greg Lynn and Daniel Libeskind, often make references to nature and to the natural sciences, although their work can be seen as a re-interpretation if not a challenge of the organic paradigm.

In most cases, such a design approach was based more on semiotic than aesthetic concerns. The champions of *l'architecture parlante*, for example, reached for a universally understandable language of architecture. Makovecz, on the other hand, has attempted to get to the essence of architecture by investigating the biomorphic etymologies of Hungarian words related to buildings. Other architects have sometimes adopted natural forms for structural purposes. Henrik Petrus Berlage, inspired by one plate of Ernst Häckel's popular *Art Forms in Nature*, used one of his jelly fish images to design a lamp (Fig. 8.2).

However, the more recent designs of Frank Gehry, Future Systems, Renzo Piano, and others also show strong affinities to the structures of natural organisms. Santiago Calatrava and Nicholas Grimshaw, for instance, design expressive skeletons that fold and bend like body parts, as can be seen in Calatrava's garage door that folds like an eyelid. Even more explicit may be the shape of the H_2O Pavillon by Nox and Oosterhuis architects, which evokes a stranded whale while its interior

8 Biomorphism in Architecture: Speculations on Growth and Form 153

Fig. 8.2 Ernst Haeckel, Acanthrophracta- Tafel aus Kunstformen der Natur

seems to address the fluidity of a virtual, liquid "Deleuzian" space.[1] The computer-generated designs of Greg Lynn, Lars Spuybroek, and Jeff Kipnis follow a "blob grammar," which resembles the amorphous forms and viscous transformations of natural organisms one sees in floating jellyfish [7].

8.4 Natural Processes

In his Hydrogen House project for Vienna, Greg Lynn for example started with a simple, symmetrical triangular volume which gets deformed in a process by solar rays and the shadows they cast on the building. Lynn calculated each phase of the transformational process with a computer and simulated the changing shape

---

[1] Deleuze considers traditional notions of space and time as unifying forms imposed by the subject. He claims that pure difference is non-spatio-temporal, an idea, which he calls "the virtual".

Fig. 8.3 Greg Lynn, Hydrogen House Vienna, credit: © Greg Lynn

of the building (Fig. 8.3). The result is a metamorphosis of form as a movement in time proceeding from east to west. In this regard, the Hydrogen House no longer confesses to "the ethics of stasis". Lynn explains: "Architectural form is conventionally conceived in a dimensional space of idealized stasis, defined by Cartesian fixed-point coordinates. An object defined as a vector whose trajectory is relative to other objects, forces, fields and flows, defines form within an active space of force and motion. This shift from a passive space of static coordinates to an active space of interactions implies a move from autonomous purity to contextual specificity" [8].

For the Hydrogen House and other designs, Lynn has produced computer videos that explain the generation of the forms from the dynamic interaction of contextual forces. The animated sequence illustrates the dynamism vividly, but at some point, Lynn always stops the process and chooses a static image as the finished design for the building is not itself supposed to move. Many critics have complained about Lynn fetishizing the dynamic animation video but then taking the motion away from the actual architectural object. Yet it is possible that the Hydrogen House is not the best illustration of animate form: because the work was for a while intended to be built, Lynn might have had to make compromises concerning construction, cost, or other factors. His Long Island House Project, on the other hand, must be taken as representative of his main interests, as no external matters enter into the design process. It is remarkable how closely the project resembles late sculptures by Umberto Boccioni.[2]

---

[2] The design began as an analyses for a small weekend house. Lynn mapped the site based on visual attractors using forces of various shapes. Into this field of forces he placed various flexible house prototypes in order to study their alignments and deformations.

8 Biomorphism in Architecture: Speculations on Growth and Form

Some of Lynn's designs end up much the same way as Boccioni's sculpture: as symbolic depictions of movement or dynamic forces. The objects simply are not animate, unless the word "animate" is understood in a special sense. Indeed, Lynn articulated this sense in explaining that "animation is a term that differs from, but is often confused with, motion. While motion implies movement and action, animation implies the evolution of a form and its shaping forces; it suggests animalism, animism, growth, actuation, vitality and virtuality" [9]. While Lynn's characterization of "animation" (which is probably meant to apply to "animate" as well) differs from those in standard dictionaries, it is nonetheless enlightening. Most of the words in the above quote are recognizable as shorthand for ideas that Lynn discussed in his earlier essays: "actuation" and "virtuality" referred to Deleuze's philosophy, "growth" perhaps to D'Arcy Thompson's *Growth and Form*, and "vitality" probably to Bergson's vitalism [10]. The word "animism" seems to be a recognition of the fact that seeing Lynn's designs as animate require a kind of *Einfühlung* or empathy which is also present in animistic religions. To understand Lynn's reference to "animalism", however, one needs to consider an earlier discussion of animate form: theosophical speculation on "hyperspace", in particular by P. D. Ouspensky and Claude Bragdon.

Long before Lynn, Frei Otto explored different ways of re-interpreting natural systems to develop architectural shapes and structures. Opposed to Lynn, he was interested in the form rather than the process or motion. From the early 1970s, Frei Otto began fusing forms found in nature with modern building techniques and computer logistics. His book *Biology and Building 2* (1972) examined ways in which the lightweight sandwich construction of bird skulls could be applied to architecture; a further volume published the following year dealt with the strength and beauty of spiders' webs. The goal of Otto was to stretch man-made structures to their limits with a most economical use of material. Further research examined the structure and building properties of for instance bamboo and soap bubbles (Fig. 8.4).

Fig. 8.4 Frei Otto, credit: © BauNetz Media, see www.baunetz.de/db/news/?news_id=82689

Otto observed that given a set of fixed points, soap film will spread naturally between them to offer the smallest achievable surface area. Several experiments followed and Otto established his position between architect, artist, and engineer. His quest to discover light, strong, responsive, and elegant structural solutions for buildings drew his research toward nature as reflected in his 1995 study *Pneu and Bone*, which considered the structural properties of crustaceans [11]. Another very promising direction has more recently been taken by Shigeru Ban, who was certainly influenced by Frei Otto, with whom he later also collaborated, by developing architectural shapes in relation to natural bubble and parabolic structures.

8.5 Organic Versus "Mechanical" Form

Organic forms have often been opposed to geometric forms. Peter Sloterdijk's book *Sphären* has been considered by many architects to be a standard reference work of organically shaped and pneumatic architectonics such as presented in the early 1970s by Cedric Price, Coop Himmelb(l)au or Haus-Rucker, to name but a few [14c] (Figs. 8.5 and 8.6). The idea was to create a new kind of living environment which is not only a second skin for its human inhabitants but rather an organically shaped living structure. However, under the influence of aerospace industry, such formal ideas were often connected to utopian visions of a future world with new building materials, future techniques, and of course a different society that was still to come. Reyner Banham critically described these architects as "Zoom Wave Newcomers" who present an alternative architecture – that would be perfectly possible if only the Universe was differently organized [12, 13].

Fig. 8.5 Haus Rucker & Co, credit: © Haus Rucker & Co

8 Biomorphism in Architecture: Speculations on Growth and Form

Fig. 8.6 Installation von Haus Rucker & Co am Friedericanium in Kassel zur documenta 5, credit: Haus Rucker & Co

Although Frei Ottos experiments had a much more rational starting point, they were still a challenge for the clean cut right angle of modernism (Figs. 8.7 and 8.8). Yet organic and geometrical shapes were not always seen as oppositions. Louis Sullivan, for example, started with a simple square intersected by diagonal and orthogonal axes to create delicately floral motifs. In his "Essay on Inspiration," he describes the fusion of geometric and organic forms as a design principle of nature, finding in it a transcendental, religious dimension. Influenced by Emanuel Swedenborg, a Swedish scientist, philosopher, Christian mystic, and theologian, Sullivan recognizes the "feminine" principle in floral, organic forms which emerge from the underlying geometric "masculine" form. The idea that life is generated from such oppositions and that the universe rests on a dualist foundation is a fundamental idea in his architecture [14].

Interestingly enough, Sullivan saw no contradiction between ornament and function, despite his reputation as a forerunner of modernist theory; rather, he considered ornament as a necessary element. The opposition of geometric versus amorphous forms, which was not an issue for Sullivan, bitterly divided later theorists. One problem concerned the definition of organic form. Claude Bragdon saw two fundamental possibilities, designed vs. organic architecture, and recognized this duality as a basic principle of life. Designed architecture is conceptual and

Fig. 8.7 Frei Otto, Olympic Stadion, Bilderarchiv Institut für Architekturwissenschaften, TU Wien

Fig. 8.8 Frei Otto, Olympic Stadion, Bilderarchiv Institut für Architekturwissenschaften, TU Wien

artificial, created by talent and influenced by taste, whereas organic architecture is unconscious, free, and imaginative. However, he had to concede that the two kinds of architecture could not always be clearly separated [15]. In a similar vein, Walter Curt Behrendt talked in the 1930s about the opposition between organic and mechanical order [16].

Numerous attempts to limit organic theory to the formal phenomenon alone could not survive closer scrutiny and in practice led to an incoherent set of divergent variations. Antoni Gaudi's or Hermann Finsterlin's positions, for example, were criticized by Bruno Zevi who felt that organic architecture should never be understood as the application of forms derived from or inspired by plants and animals nor as the more metaphorical representation of nature. Again and again in his *Towards an Organic Architecture*, he disavows the use of biomorphic imagery which in his mind reduces aesthetic pleasure to physiological or sexual sensations [17].

A new solution to the age-old dilemma of organic vs. mechanical form was offered in the 1970s by the popularization of fractal geometry. This branch of mathematics was popularized by Benoit Mandelbrot, who promoted it as a geometry of nature. Frustrated with the inadequacy of mathematics to model certain natural phenomena, Mandelbrot found that the apparent disorder of nature reveals, on closer inspection, repetitions of certain structures [18]. He was able to provide equations that reproduced the irregular, fragmented patterns of natural phenomena, through the use of iterative processes: branches of trees made from innumerable smaller branchlets, a convoluted coastal landscape comprised of countless smaller involutions, the shape of a feather created out of myriad smaller "feathers" at ever smaller scales. Mandelbrot traces this observation back to Eugene Delacroix who in his turn refers to Swedenborg's claim "that the lungs are composed of a number of little lungs, the liver of little livers, the spleen of little spleens" and so forth [19].

8.6 Bionics and Cyborgs

The American author and architecture theorist Anthony Vidler stated in the early 1990s that the boundaries between the organic and the inorganic have been blurred by cybernetic and biotechnologies. Nowadays, they seem to be less sharp than ever before: "the body, itself invaded and re-shaped by technology, invades and permeates the space outside, even as this space takes on dimensions that themselves confuse the inner and the outer, visually, mentally and physically" [20].

Indeed, inspired by countless science fiction novels, recent developments in medicine and virtual intelligence, intelligent materials, automation, etc., the architectural discourse turned once more to the old idea of the house as second skin or living being.[3] However, this time the question was where to draw the line between

---

[3]This debate originated in Jack E. Steeles work, who coined the term bionics, and his research on cyborgs, which inspired Science Fiction writer and aviation expert Martin Caidin, to write his famous Cyborg book in 1972. The popular American TV series The Six Million Dollar Man (and the following spinoff The Bionic Woman), popularized, unfortunately somewhat inaccurately, the term bionics. Steele's original intention for bionics was the study of biological organisms to find solutions to engineering problems, a field which is today known as biomimetics. The interpretation of the biological paradigm represented by the so-called signs of life was examined in [21].

building and user or the natural and the artificial. Donna Haraway's famous *Cyborg Manifesto* became a strikingly important new philosophy for several architects and theorists which sponsored again an interest in natural processes and the question of what the essence of an organism might be: "A cyborg is a cybernetic organism, a hybrid of machine and organism, a creature of social reality as well as a creature of fiction. Social reality is lived social relations, our most important political construction, a world-changing fiction. The international women's movements have constructed 'women's experience', as well as uncovered or discovered this crucial collective object. This experience is a fiction and fact of the most crucial, political kind. Liberation rests on the construction of the consciousness, the imaginative apprehension, of oppression, and so of possibility. The cyborg is a matter of fiction and lived experience that changes what counts as women's experience in the late twentieth century. This is a struggle over life and death, but the boundary between science fiction and social reality is an optical illusion" [22]. As Haraway pointed out, cyborg refers to cybernetic organism, describing a kind of hybrid of the natural and the artificial, mechanical. Originally, the term "cybernetics" was introduced by the mathematician Norbert Wiener in the 1940s to describe the study of complex systems of control and communications in animals and machines. Hernan Diaz Alonso explored these issues in several formal experiments, which would usually turn into exhibition architecture but sometimes housing projects as well. His designs recall the aesthetics of science fiction artist Giger yet also surpass the borderline between art and architecture, sculpture and building, organism and machine (Figs. 8.9–8.11).

However, soon the debate moved from the individual building to the urban context and contemporary city planning including infrastructure, as Gandy pointed

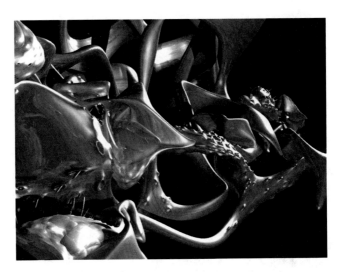

Fig. 8.9 Hernan Diaz Alonso, Prototype Concepts Cell Phone, credit: *Diaz-Alonso Project: Cell 2006*/Cell Phone/Prototype Concepts Client – Confidential © Xefirotarch/MAK

8 Biomorphism in Architecture: Speculations on Growth and Form 161

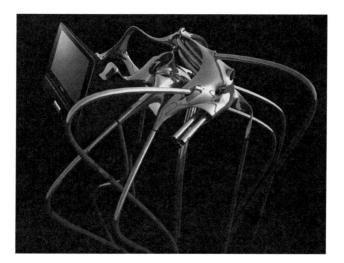

Fig. 8.10 Hernan Diaz Alonso, *Exhbtn Detail Pitch Black, "Spider"*, credit: *Diaz-Alonso PITCH-BLACK, 2007, Rendering, Hernan Diaz Alonso* Detail, Spider © Xefirotarch/MAK

Fig. 8.11 Hernan Diaz Alonso, *Project: Stockholm* 2006, credit: *Diaz-Alonso, Project: Stockholm* 2006/Stockholm/Competition, Stockholm Sweden, © Xefirotarch/MAK

out: "The emphasis of the cyborg on the material interface between the body and the city is perhaps most strikingly manifested in the physical infrastructure that links the human body to vast technological networks. If we understand the cyborg to be a cybernetic creation, a hybrid of machine and organism, then urban infrastructures can be conceptualized as a series of interconnecting life-support systems" [23]. The modern home, for example, has become a complex exoskeleton for the human body with its provision of water, warmth, light, and other essential needs. The home can be conceived as "prosthesis and prophylactic" in which modernist distinctions between nature and culture, and between the organic and the inorganic, become blurred. And beyond the boundaries of the home itself we find a vast interlinked system of networks, pipes and wires that enable the modern city to function. These interstitial spaces of connectivity within individual buildings extend through urban space to produce a multi-layered structure of extraordinary complexity and utility" [24].

From this perspective, it is only a short move to questions of superorganisms, ecology, and more recently new concepts of sustainability, and the idea, that everything is connected, while a small change in one area might challenge the whole system (as exemplified in the butterfly effect).

8.7 Ecology

A building could be seen as issuing from its inhabitant, but it could also be seen as growing from the earth like a plant, much as national styles and art forms (according to Johann Gottfried Herder) arose from the soil of their time and place. In recent years, ecological architecture has become the focus of much research, provoked by the 1970s energy crises. However, while attempts to develop sustainable architectural systems with minimal energy use and minimal waste are certainly important, the theoretical implications of ecology also need to be addressed. One of the central issues concerns the individuation of organisms in ecological thought. Instead of conceiving of a plant or an animal as a separate entity as would have Carl Linneus, ecologists usually focus on populations and the interrelation of organisms and nonliving nature in ecosystems. A squirrel could in fact not exist without the plants it eats; these would not grow except for certain minerals, water, air, etc. Logically applied, then, the ecological point of view entails the concept of an ecological *superorganism*, a concept proposed by Frederic Clements in 1916. It states that different ecosystems are organisms in their own right, with particular emergent properties that their constituent parts, animals, and plants do not have.

The application of such considerations to architectural and urban design raises several questions. Just as no organism is self-sufficient but rather merely an open system interacting with others in a larger ecosystem, neither are buildings self-sufficient nor independent. In cities, building tap into the infrastructure of water pipes and sewers, electric lines and communications, and streets. This idea was given artistic and graphic expression in Peter Eisenman's *Wexner Center for the Visual Arts* in Cincinnati, where the grid of the building extends onto the sidewalks as inlaid brick in the concrete.

Frei Otto and Shigeru Ban addressed this question of interconnection and interaction of architectural systems and their environment from a global ecological perspective. The main theme of their Japan Pavilion at Hanover Expo was to create a structure that would produce as little industrial waste as possible when it was

8 Biomorphism in Architecture: Speculations on Growth and Form

Fig. 8.12 Frei Otto, Shigeru Ban, Japanese Pavillon, Expo Hanover, Model, credit: shigeru ban architects

Fig. 8.13 Frei Otto, Shigeru Ban, Japanese Pavillon, Expo Hanover, exterior, credit: shigeru ban architects

dismantled (Figs. 8.12 and 8.13). The goal was either to recycle or to reuse almost all of the materials that went into the building. The structural idea is a grid shell using lengthy paper tubing without joints. The tunnel arch was about 73.8 m long, 25 m wide, and 15.9 m high. The most critical factor was lateral strain along the long side; so instead of a simple arch, a grid shell of three-dimensional curved lines was chosen with indentations in the height and width directions, which are

Fig. 8.14 Shigeru Ban, Modell Paper-Log-House, Kobe, credit: shigeru ban architects

stronger when it comes to lateral strain. One has to keep in mind that the artistic agenda behind this project is just as important as in Eisenman's building, the main difference being that Otto and Ban preferred nonstandard, innovative structures and materials, while considering issues of industrial waste.

Shigeru Ban's temporary "log" houses in Kobe, Kaynasli, and Bhuj also explore issues of ecology and sustainability, however, this time built for the victims of the earthquakes in catastrophe areas (Fig. 8.14). The temporary shelters have foundations that consist of donated beer crates loaded with sandbags. The walls are made of 106-mm diameter, 4-mm thick paper tubes, with membrane roofs. For insulation, a waterproof sponge tape backed with adhesive was sandwiched between the paper tubes of the walls. Sustainability does not only concern building materials but also concern social issues.[4] The units are easy to dismantle, and the mostly local materials can be easily disposed or recycled. The log houses in India show a particular feature. It was coated with a traditional mud floor. For the roof, split bamboo was applied to the rib vaults and whole bamboo to the ridge beams. A locally woven cane mat was placed over the bamboo ribs, followed by a clear plastic tarpaulin to protect against rain, then another cane mat. Ventilation was provided through the gables, where small holes in the mats allowed air to circulate. This ventilation also allowed cooking to be done inside, with the added benefit of repelling mosquitoes.

---

[4] This building method proved to be very cheap and therefore affordable: the cost of materials for one 52 square meter unit was below $2000.

8.8 From Fractals to Catastrophies

Eisenman and his followers, however, are interested not so much in practicing ecology but in processes as discussed in natural sciences, mathematics, and physics. In the 1980s, Eisenman aspired to a "tectonic literature" that will write itself, absolving him from the responsibility of authorship and the guilt of authority. To this effect, he proposed several design methods that attempted to dislocate the author from the work by replacing the designer's intentional choice with either aleatory systems or the impersonal determinism of an algorithm. A number of his projects, built and unrealized, were developed using scientific models: fractals in the *Wexner* Center in Columbus Ohio, 1989; the genome in the Biozentrum in Frankfurt, 1989; "Boolean" cubes for the Carnegie Mellon Research Institute in Pittsburgh; the "butterfly cusp" diagram of catastrophic events for the Rebstock Park Housing in Frankfurt, 1992; the Möbius strip for the Max Reinhardt Haus project for Berlin, 1993; and soliton wave studies for the Jörg Immendorff Haus in Duesseldorf, 1993.

The longest standing of these was his interest in the "scaling" aspect of fractal geometry. Borrowing the term from Mandelbrot, but ignoring the mathematician's intuition that "the fractal new geometric art shows surprising kinship to Great Master paintings or Beaux Arts architecture," Eisenman attempted to revive the avantgarde project by reading fractals as analogous to Jacques Derrida's deconstruction, and vice versa [25]. Eisenman believed that fractal geometry could overturn the "metaphysics of scale" in Western architecture, because the self-similarity of fractals destroyed the possibility of originality, an originary trace, and a decidedly "real" scale [26]. This interpretation must, however, be rejected. If self-similar recursivity questions origin, so too do self-sameness, reflexivity, resemblance, or similitude in general, and fractals have no monopoly on dislocating origins. Furthermore, in Eisenman's projects such as the Wexner Center or Frankfurt Biozentrum, self-similarity was limited to but a few privileged points, while in fractal geometry, every point has that property. Finally, Eisenman introduced neither self-similarity nor the principle of scaling to architectural design. Charles Jencks points out that Bruce Goff "virtually invented fractalian architecture before the fact" in his design for the studio of Joe Price, overlaying of triangles, hexagons, trihexes, and lozenges [27]. In fact, the repetition of self-similar forms is found throughout architectural history, most evident in Gothic cathedrals and even in Greek and Egyptian temples.

Although scaling, folding, and other mathematically inspired design algorithms fail to bear out authorial claims about their fractal, deconstructive, rhizomatic, or scientific nature, they do make some progress toward realizing the ideal, proposed by both Hugo Häring and Stéphane Mallarmé (among countless others), of completely erasing the author and letting the work create itself in a natural, unmediated way [28]. Influenced by Hegelian philosophy, Mallarmé wrote to Henri Cazalis already in 1867, "I am impersonal now: not the Stéphane you once knew, but one of the ways the Spiritual Universe has found to see Itself, unfold Itself through what used to be me" [29].

In some of Eisenman's designs, such as his *Romeo and Juliet* project of 1986, the designer almost disappears, just letting various discourses, including Shakespeare's play and the city of Verona, to intersect and unfold in new combinations. Completed for the 1986 Venice Biennale, the project was to present the dominant themes of the Romeo and Juliet story in architectural form. Through the diagrammatic processes of superposition and scaling, the plan of the city of Verona was transformed to reveal the themes of Shakespeare's play.

While Eisenman's algorithmic designs have succeeded at times in displacing the author and disrupting the standard categories of architecture, his methods have been criticized by his own authority. Jacques Derrida accused Eisenman's scaling of being "totalizing first because it is structured as a closed narrative entirely determined by origin and end – it does not respect textual openness and indeterminacy. Secondly, scaling is an anathema because it is the vehicle by which Eisenman seeks to replace one totality, traditional design, with a new and different totality" [30].

Chastened, Eisenman and his followers moved from Mandelbrot's fractal geometry and Derrida's deconstruction to René Thom's catastrophe theory paired up with the philosophical ideas of Gilles Deleuze. In its emphasis on unpredictable, locally emergent properties which result from feedback loops and irreducible, nondecomposable diffusion of decisive factors, catastrophe theory seemed the ideal response to Derrida's charge of totalization in the scaling method. Inspired by D'Arcy Wentworth Thompson, René Thom had developed catastrophe theory as a way of addressing biological morphogenesis mathematically. Catastrophe theory can supply a number of descriptions of morphological transformations – provided they can be approximated by dynamic systems with fixed points as their attractors – but mathematicians are divided over the means of deciding whether observable discontinuity on the level of phenomena can be interpreted as a mathematical jump in the space of the attractors. However, Peter Eisenman's Max Reinhardt Haus project for Berlin (1993), which is partly based on these ideas, resembles the crystalline architecture of the Czech Cubist Pavel Janak.

The point has often been made that the architecture of Eisenman, Lynn, Chu, NOX, Marcos Novak, and others bears a certain formal similarity, for example, to the *Merzbau* of Kurt Schwitters, to Czech Cubism, to Boccioni's futurism, and to the chronophotographs of Etienne-Jules Marey. One of the reasons why the designs of the new avantgarde look so much like experiments of a 100 years ago might be that contemporary architects are still reading the same books, looking at the same pictures, and engaging in the same issues as the artists of around 1900.

Although most applications of catastrophe theory (by Christopher Zeeman and others) have produced empirically false predictions, this does not necessarily invalidate it as an explanatory paradigm. However, the application of catastrophe theory in architecture is a different ball game altogether, since we are dealing not with descriptive but normative issues. Robin Evans once suggested that Peter Eisenman's houses were not so much *analogs* to language as they were three-dimensional models of Noam Chomsky's *theory* of language. Eisenman's work in the 1980s can similarly be seen as attempts to petrify Jacques Derrida's beliefs about trace, *différance* and effacement in concrete, steel, and glass [31]. His more recent applications of the work of Deleuze or Thom are not essentially different – the question is why these theories about natural or social processes, which they purport to describe, should be reproduced in architecture?

Different reasons have been proposed by contemporary writers. Greg Lynn suggests that architects should use complex curved and folded planes because recent advances in computer modeling have made topological descriptions of such forms accessible to non-mathematicians [28]. Why the mere possibility of drafting certain forms with precision would justify those forms is, however, unclear – unless one subscribes to a variation of the classic principle of plenitude (i.e., the thesis that all true potentialities need to be actualized) or to some kind of organicist belief that everything is interconnected and hence every innovation must have consequences in every domain of life.

A different defense for the use of chaos theory in architectural design is provided by Charles Jencks, who claims that the task of architecture is to tie human beings into the cosmos by building close to nature, thereby representing "the basic cosmogenic truth" of self-organization, emergence, and jumps to a higher level. Like Renaissance theorists, he insists that architecture must look to contemporary science for "disclosures of the Cosmic Code" and claims that Frank Gehry's design for the Guggenheim Museum at Bilbao reflects best the new paradigm. Only by looking "to the transcendent laws which science reveals," can architecture "get beyond the provincial concerns of the moment, beyond anthropomorphism and fashion" and "regain a power that all architecture has had" [32]. While ostensibly declaring avantgarde architecture to be as advanced and intellectually respectable as modern science, Jencks' New Age vision nevertheless relegates architecture to an inferior position, subordinate to natural science.

8.9 Form Follows Function

In order to fulfill the conditions of an organic unity or organic whole, it has been common in western architecture to make use of the proportions of natural beings rather than reproducing their forms. Vitruvius advocated basing the proportions of a building on those of a perfect man, establishing a tradition that inspired numerous reconstructions and revisions, the most famous one in modern architecture being Le Corbusier's *Modulor* of 1948, which overlaid the image of a man on the Fibonacci series, found in the shell of the spiral nautilus and which formed the basis of the golden section.

Yet in the eighteenth century, British empiricists had argued against the idea that architecture should imitate the proportions of natural organisms. In *A Philosophical Enquiry into the Origin of Our Ideas of the Sublime and Beautiful*, Edmund Burke argued that proportion is not the cause of beauty in vegetables, nor can the notion of architectural proportion be derived from the Vitruvian man. His major arguments against the Vitruvian doctrine ran as follows: "Men are very rarely seen in this strained posture; it is not natural to them; neither is it at all becoming. – the view of the human figure so disposed, does not naturally suggest the idea of a square, but rather of a cross; – several buildings are by no means of the form of that particular square, which are notwithstanding planned by the best architects, and produce an effect altogether as good. [Finally, Burke concluded,] no two things can have less resemblance or analogy, than a man, and a house or temple: do we need to observe, that their purposes are entirely different?" [33].

Burke's tacit assumption was that if the purposes of a man and a house are different, then their forms should be different as well. In this, he relies on another Aristotelian commonplace: that an entity is defined by its *telos* or goal. For Aristotle, *ars imitatur naturam* meant that artists should work like nature – not by imitating the appearance of natural organisms, but by letting their creations unfold their own natures. If Aristotle is correct, nature does nothing in vain, "God and nature create nothing that has not its use" [34].

The fundamental tenet of functionalism that of designing "von innen nach aussen" agrees with Aristotelian essentialism [35]. Mediated by Romantic and Transcendental thought, the Aristotelian principle of creation was reformulated in 1896 by Louis Sullivan, "It is the pervading law of all things organic, and inorganic, of all things physical and metaphysical, of all things human and all things superhuman, of all true manifestations of the head, of the heart, of the soul, that the life is recognizable in its expression, that form ever follows function. This is the law." [36]. Many architects turned directly to nature to study somewhat optimized structures, in particular load bearing systems and streamline forms like so many projects by Buckminster Fuller (Dymaxion Car 1933, Dymaxion House 1945, Geodesic Dome in Montreal 1967) (Figs. 8.15–8.18). Tree structures and tent structures were also objects of investigation among modern architects. With his students, Frei Otto developed already in the early 1960 various concepts of branched pillars after the modell of trees. The feasibility study "Tree Structures" for an exhibition hall at Yale University, USA, in 1960 was further developed in the support pillars of a six-angle gridshell in the Kings Office at the Council of Ministers in Majlis al Shura, Riyadh, Saudi Arabia, 1979 (Frei Otto together with Rolf Gutbrod, Büro Happold, Ove Arup, and Partner).

Santiago Calatrava's designs are often inspired by nature, featuring a combination of organic forms and technological innovation. A good example is the Milwaukee Art Museum expansion, which incorporates multiple elements inspired by the Museum's lakefront location (Figs. 8.19 and 8.20). Among the maritime elements in Calatrava's design can be distinguished movable steel louvers inspired by the wings of a bird; a cabled pedestrian bridge with a soaring mast inspired by the form of a sailboat and a spider web and a curving single-storey galleria reminiscent of a skeleton or a wave. The smooth flight of a giant bird inspired the spectacular extension for the main building whose immense "wings" open and close with the museum. More or less the same form is repeated in his vision for a Port Authority transit hub in New York City (Fig. 8.21). A comparable bird-like structure seems to be also the basis of the Qatar Photography Museum, which opens and closes

8 Biomorphism in Architecture: Speculations on Growth and Form 169

Fig. 8.15 Buckminster Fuller, Dome and Car, Bilderarchiv Institut für Architekturwissenschaften, TU Wien

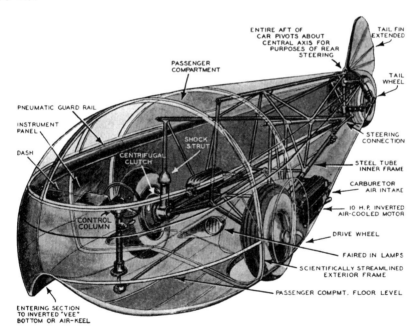

Fig. 8.16 Buckminster Fuller, Dymaxion Car, Bilderarchiv Institut für Architekturwissenschaften, TU Wien

Fig. 8.17 Buckminster Fuller, Geodesic Dome, Bilderarchiv Institut für Architekturwissenschaften, TU Wien

Fig. 8.18 Ernst Haeckel, Detail from Art Forms in Nature, Bilderarchiv Institut für Architekturwissenschaften, TU Wien

depending on the amount of sunlight. The key idea of the Museum of Photography is an ultra-light structure consisting of two immense curved "wings" which will open and close with the light, fine-tuned to the needs of the exhibitions, just like the lens of a camera, which goes back to the movement of the pupil.

Fig. 8.19 Santiago Calatrava, Art Museum Milwaukee, © calatrava architects

8.10 The Concept of Organic Unity

In architectural theory, the notion of *organic unity* was accepted on Aristotle's authority until the nineteenth century, when new interpretations were proposed on the basis of contemporary biology. Paraphrasing the biologist Georges Cuvier, the architect Eugène-Emmanuel Viollet-le-Duc wrote: "Just as when seeing the leaf of a plant, one deduces from it the whole plant; from the bone of an animal, the whole animal; so from seeing a cross-section one deduces the architectural members; and from the members, the whole monument" [37].

This argument was later repeated by Adolf Loos, who claimed that one could reconstruct an entire society from a single button.

Both Gottfried Semper and Viollet-le-Duc turned to biology to argue their position in the debate as to whether form follows function or vice versa. This debate began in the 1830s with the establishment of morphology and comparative anatomy. While Cuvier proposed the "form follows function" theory, his rival Etienne Geoffroy Saint-Hilaire insisted one could not draw conclusions for the structure by looking at the function, arguing that no matter what their function, all organic forms could be reduced back to original types. In this conviction, he came close to Goethes earlier theory of the *Urpflanze*, the original plant.

Fig. 8.20 Santiago Calatrava, Art Museum Milwaukee, interior, Bilderarchiv Institut für Architekturwissenschaften, TU Wien

By the 1840s, biologists rejected both Cuvier's and Saint-Hilaire's position and questioned in general the validity of teleological interpretations of natural phenomena. Contrary to the theses of Viollet-le-Duc and Cuvier, an investigation of any organism immediately reveals an arbitrary number of parts, shape, structure, or function of which allow for no adequate explanation. Yet, despite the efforts of a century of Darwinism to present evolution as a process involving the random mutation of genes, with natural selection weeding out only those mutations that fatally affect populations, many organicist writers continue to view evolution as a teleological process of improvement in which organisms have achieved a perfect adaptation to their environment – permitting some ecologists to take it as axiomatic that all ecosystems are in perfect homeostasis, only upset by the thoughtless interventions of humankind. Such an idea of nature, as a perfectly functioning ecological complex, is not derived from empirical observation but rather from extra-scientific, usually theological, sources.

Fig. 8.21 Santiago Calatrava, NYC, Transportation Hub, © Courtesy of the Port Authority of New York and New Jersey

Be that as it may, the Aristotelian ideal of an organic whole has pervaded modern architecture, even the theories of architects who never used natural forms in their designs. Thus, Ludwig Hilberseimer, for example, demanded that "all works, however different, must originate in a unified spirit" [38]. Similarly, Ludwig Mies van der Rohe defining a structure turns to the Aristotelian concept of organic wholes when he says, "by structure we have a philosophical idea. The structure is the whole, from top to bottom, to the last detail with the same ideas" [39].

In what is perhaps the most virulent formulation of organicism, the philosophy of G.W.F. Hegel, the concept is understood in terms of the interconnectedness or the essential fragmentary nature of everything as Richard Shusterman pointed out: In his *Logic*, Hegel wrote: "Everything that exists stands in correlation, and this correlation is the veritable nature of every existence. The existent thing in this way has no being of its own, but only in something else" [40, 41].

According to the British philosopher George Edward Moore, the Hegelian notion of an organic unity claims that "just as the whole would not be what it is but for the existence of its parts, so the parts would not be what they are but for the existence of the whole" [41, 42]. The idea is that emergent properties belong to the parts as well as the whole. A severed hand, for example, cannot function at all in the way we expect of a hand. Things with different properties must be different things, so that a hand severed from its body detached is not the same thing as a hand still connected.

Hence, parts of an organic whole (in the Hegelian sense) are inconceivable, except as parts of that whole.

Yet Moore rejected this view, because it confuses properties belonging to the whole with properties of one of its parts. Even more significantly, he viewed the Hegelian notion of an organic whole as self-contradictory, because it assumes that a part is distinguishable from the whole while simultaneously asserting that the part contains some aspects of that whole as part of itself. In other words, radical organicism postulates that any individual part we distinguish as an element in the whole cannot be so distinguished – the part *is*, it is not part of a whole.

Obviously, Derrida's theories are a form of radical organicism, based on the universalization of Ferdinand de Saussure's diacritical linguistics or Friedrich Nietzsche's belief that "in the actual world [...] everything is bound to and conditioned by everything else" and his conclusion that "no things remain but only dynamic quanta, in a relation of tension to all other dynamic quanta: their essence lies in their relation to all other quanta" [41, 43].

We have already seen that radical ecology makes similar claims for biological systems. In Derrida's hands, organic diacriticism leads, among other things, to the dissolution of the work of art as an independent entity, as defined by modernist criticism.

Hence, the very concept of an organic unity is self-contradictory. Yet, even poststructuralist writers accept the notion in their critical practice, for the assumption of unity is required of the concept of a work of art which in turn forms the basis of the interpretive activity [44–46]. The concept of organic unity is therefore inexorably bound with the concept of a work of art – only if taken as a unity does a thing exist as a work of art [47]. In the same way, Mies van der Rohe compared different types of buildings to roses and potatoes, explaining that, "while both are based on the same natural principles, we ask of a rose only that it be a rose; we ask of a potato only that it be a potato. Philosophically speaking, only then do they exist." [48].

8.11 Conclusion

Despite their authors' claims of producing something radically new, many of the design strategies applied by the current architectural avantgarde can be traced back to one of the oldest and most influential ideas in architectural history: the concept of organicism in its various guises. Aristotle's definition of art as the imitation of nature provided the unquestionable premise for two millennia of architecture.

While attempts to base architectural design on the example of nature are typical of stylistic crises and indicative of a search for new, possibly objective foundations, the implications of organicism are not limited to particular styles or forms [49–51]. Rather, they are general and fundamental enough to relate to contemporary issues at any time and pose radical questions. More recently, the challenge of *ars imitatur naturam* focuses on the author and the object, sustainability, and imitation of natural processes.

The objective of following "nature" as revealed through the natural sciences or the "nature" (essence) of the particular building de-emphasizes the authorial self-expression of the architect and ultimately leads to the "death of the author," as announced by Roland Barthes in the artistic context or the reconstruction of natural materials and processes within the structural context. The notion of a building as a functional unity comparable to an organism necessitates a questioning, on one hand, of what is a function, how it is historically, physically and socially constituted, and what would qualify as a functional unit of study, and on the other hand, how body, function, and building and the chiasmatic relationships between them may be articulated. If the argument sketched above is correct, the assumption of the building as an autonomous entity or an independent whole has soon to be given up.

References

1. D.W. Thompson, *On Growth and Form* (Cambridge University Press, Cambridge, 1942), p. 16
2. Aristotle, *De Partibus Animalum* and *De generatione* (Clarendon, Oxford, 1972) (407b1; cf. *Pol.* 1325b10)
3. Aristotle, *Ethica Nicomachea (Nicomachean Ethics)*, trans. with commentaries and glossary ed. by H.G. Apostle (Peripatetic Press, Grinnell, 1984), p. 1168a7 and S.R.L. Clark, *Aristotle's Man. Speculations upon Aristotelian anthropology* (Clarendon Press, Oxford, 1975), p. 62
4. G. Leopardi, "A se stesso." The translation is Jormakka's; the original is as follows: "Omai disprezza/Te, la natura, il brutto/Poter che, ascoso, a comun danno impera/E l'infinita vanità del tutto," in *Selected Prose and Poems*, ed. by I. Origo, J. Heath-Stubbs (Oxford University Press, London, 1966), p. 280
5. C. Thacker, *The Wildness Pleases. The Origins of Romanticism* (Croom Helm, London and Canberra; St. Martin's Press, New York, 1983), p. 88
6. Le Corbusier, *L'art decoratif d'aujourd'hui*, éd. by Arthaud (Paris, 1980), p. 176
7. K. Jormakka, D. Kuhlmann, O. Schuerer, *Design Methods (Basics)* (Birkhäuser, Basel, 2007), p. 9ff
8. G. Lynn, An Advanced Form of Movement, in *Architectural Design*, vol. 67 (1997), p. 54ff
9. G. Lynn, Animate form. *Faultlines*. UCLA Department of Architecture and Urban Design. 1997, vol. 1, No. 2, p. 1ff
10. H. Bergson, Creative evolution (L'Evolution créatrice) (1907)
11. F. Otto, Biology and Building 2. Practical application of analogous research. From hydrostatic skeleton to skeletal muscle system. Sandwich structures within bird skulls. The principles of lightweight structures in organisms. Static of extremities. IL 4, University of Stuttgart, Germany, Otto, Frei, (1973), Pneu and Bone IL University of Stuttgart, Germany, J. Glancey, The lightweight champion of the world. *The Guardian* (Monday 4 October 2004) (1972) http://www.guardian.co.uk/artanddesign/2004/oct/04/architecture
12. P. Sloterdijk, *Sphären. Eine Trilogie* (Suhrkamp Verlag, Frankfurt, 2005)
13. R. Banham, Zoom wave hits architecture. New Soc. (1966)
14. L.H. Sullivan, *Kindergarten Chats and other Writings* (Dover Publications, New York, 1979), p. 188
15. D. Sharp, Organische Architektur — eine Art, Natur zu verstehen und wiederzugeben? *Archithese* **5**, 17 (1988)
16. W.C. Behrendt, *Modern Building* (Harcourt, Brace and Company, New York, 1937), p. 11f
17. B. Zevi, *Towards an Organic Architecture* (Faber & Faber, London, 1950)
18. B. Mandelbrot, *The Fractal Geometry of Nature* (W.H. Freeman, San Francisco, 1982)

19. B. Mandelbrot, Fractal as a morphology of the amorphous, in *Fractal Landscape from the real World*, ed. by B. Hirst (Cornerhouse, Manchester, 1994)
20. A. Vidler, Homes for cyborgs: domestic prostheses from Salvador Dali to Diller and Scofidio. Ottagono **96**, 37–38 (1990)
21. P. Gruber, *Biomimetics in Architecture [Architekturbionik] – the architecture of life and buildings*, Doctoral thesis, Vienna University of Technology, 2008
22. D. Haraway, A cyborg manifesto: science, technology, and socialist-feminism in the late twentieth century, in *Simians, Cyborgs and Women: The Reinvention of Nature* (Routledge, New York, 1991), p. 149
23. M. Gandy, Cyborg urbanization: complexity and monstrosity in the contemporary city. Int. J. Urban Regional Res. **29**(1), 28 (2005)
24. A. Vidler, *The Architectural Uncanny, Essays in the Modern Unhomely* (MIT Press, Cambridge, 1992), p. 147
25. B. Mandelbrot, Fractal as a morphology of the amorphous, in *Fractal Landscape from the real World*, ed. by B. Hirst (Cornerhouse, Manchester, 1994)
26. P. Eisenman, *Moving Arrows, Eros, and Other Errors: An Architecture of Absence* (Architectural Association, London, 1986), p. 4
27. C. Jencks, *The Architecture of the Jumping Universe* (Academy Editions, London, 1995), p. 44
28. K. Jormakka, Total control and chance in architectural design. Acta Polytech. Sc. **99**, 35–41 (1994)
29. S. Mallarmé, in *Selected Poetry and Prose*, ed. by M.A. Caws (New Directions Books, New York, 1982), p. 87
30. J. Derrida, Why Peter Eisenman writes such good books. $A + U$ E8088, p. 114 (1988)
31. R. Evans, Not to be used for wrapping purposes. AA Files **10**(69), 69 (1985)
32. C. Jencks, *The Architecture of the Jumping Universe* (Academy Editions, London, 1995), p. 167ff
33. E. Burke, *A Philosophical Enquiry into the Origin of Our Ideas of the Sublime and Beautiful* (F.C. and J. Rivington, et al., London, 1812) pp. 164ff and 183f
34. T. Aquinas, *Commentary on Aristotle's Physics*, trans. R. J. Blackwell, intro. V. J. Bourke (Yale University Press, New Haven, 1963) (II,4; Aristotle (1993) *De Caelo* 271a35; *De Part.An.*645a23–26; 639bl9)
35. H. Häring, in *Schriften, Entwürfe, Bauten*, ed. by H. Lauterbach, J. Joedicke (Karl Krämer Verlag, Stuttgart, 1965), pp. 8 and 13f
36. L.H. Sullivan, *Kindergarten Chats and other Writings* (Dover Publications, New York, 1979), p. 208
37. E.E. Viollet-le-Duc, *Dictionnaire raisonné de l'architecture française du XIe au XVIe siècle* (A. Morel, Paris, 1875–1882)
38. L. Hilberseimer, *Groszstadtarchitektur* (Hoffmann, Stuttgart, 1978), p. 99ff
39. P. Carter, Mies van der Rohe. An appreciation on the occasion this month, of his 75th birthday. Arch. Des. **1**, 97 (1961)
40. G.W.F. Hegel, *Hegel's Logic*, trans. W. Wallace (Oxford University Press, Oxford, 1975), p. 191
41. R. Shusterman, *Pragmatist Aesthetics. Living Beauty, Rethinking Art* (Blackwell, Oxford, 1992), p. 72
42. G.E. Moore, *Principia Ethica* (Cambridge University Press, Cambridge, 1959), p. 33
43. F. Nietzsche, *The Will to Power* (Vintage Books, New York, 1968) Sect. 559, 584, 635
44. G. Dickie, *Evaluating Art* (Temple University Press, Philadelphia, 1988), pp. 164–165
45. W.K. Wimsatt, M.C. Beardsley, The intentional fallacy, in *Critical Theory Since Plato*, ed. by H. Adams (Harcourt Brace Jovanovich, New York, 1971), p. 1015
46. A.C. Danto, *Beyond the Brillo Box* (Farrar Straus Giroux, New York, 1992), p. 134
47. K. Jormakka, How to read (Anything as) art. Nordic J. Arch. Res., **7**(2), 38–45 (1994)
48. F. Neumeyer, *The Artless Word. Mies van der Rohe on the Building Art*, trans. M. Jarzombek (MIT Press, Cambridge, 1991), p. 299

49. C. van Eck, *Organicism in 19th Century Architecture* (Architectura & Natura Press, Amsterdam, 1994)
50. D. Kuhlmann, *Lebendige Architektur* (Bauhaus Universität Weimar Verlag, Weimar, 1998)
51. D. Kuhlmann, K. Jormakka, *Ars Imitatur Naturam*, Thesis, vol. 6 (Bauhaus Universität Weimar Verlag, Weimar, 1997), pp. 6–13

Literature

T. Aquinas, *Commentary on Aristotle's Physics*, trans. R. J. Blackwell, intro. V. J. Bourke (Yale University Press, New Haven, 1963)
Aristotle, *De Partibus Animalum* and *De generatione* (Clarendon, Oxford, 1972)
Aristotle, *Ethica Nicomachea (Nicomachean Ethics)*, trans. with commentaries and glossary ed. by H.G. Apostle (Peripatetic Press, Grinnell, 1984)
Aristotle, *De Caelo. (On the heavens)*, trans. W.K.C. Guthrie (William Heinemann, London, 1993)
W.C. Behrendt, *Modern Building* (Harcourt, Brace and Company, New York, 1937)
E. Burke, *A Philosophical Enquiry into the Origin of Our Ideas of the Sublime and Beautiful* (F.C. and J. Rivington, et al., London, 1812)
P. Carter, Mies van der Rohe. An appreciation on the occasion this month, of his 75th birthday. Arch. Des. **1**, 97 (1961)
S.R.L. Clark, *Aristotle's Man. Speculations upon Aristotelian anthropology* (Clarendon Press, Oxford, 1975)
A.C. Danto, *Beyond the Brillo Box* (Farrar Straus Giroux, New York, 1992)
J. Derrida, Why Peter Eisenman writes such good books. *A + U* E8088, p. 114 (1988)
G. Dickie, *Evaluating Art* (Temple University Press, Philadelphia, 1988)
C. van Eck, *Organicism in 19th Century Architecture* (Architectura & Natura Press, Amsterdam, 1994)
P. Eisenman, *Moving Arrows, Eros, and Other Errors: An Architecture of Absence* (Architectural Association, London, 1986)
R. Evans, Not to be used for wrapping purposes. *AA Files* **10**(69), 69 (1985)
M. Gandy, Cyborg urbanization: complexity and monstrosity in the contemporary city. Int. J. Urban Regional Res. **29**(1), 26–49 (2005)
J. Glancey, The lightweight champion of the world. *The Guardian* (Monday 4 October 2004). http://www.guardian.co.uk/artanddesign/2004/oct/04/architecture
H. Häring, in *Schriften, Entwürfe, Bauten*, ed. by H. Lauterbach, J. Joedicke (Karl Krämer Verlag, Stuttgart, 1965)
D. Haraway, A cyborg manifesto: science, technology, and socialist-feminism in the late twentieth century, in *Simians, Cyborgs and Women: The Reinvention of Nature* (Routledge, New York, 1991)
G.W.F. Hegel, *Hegel's Logic*, trans. W. Wallace (Oxford University Press, Oxford, 1975)
L. Hilberseimer, *Groszstadtarchitektur* (Hoffmann, Stuttgart, 1978)
C. Jencks, *The Architecture of the Jumping Universe* (Academy Editions, London, 1995)
K. Jormakka, Total control and chance in architectural design. Acta Polytech. Sc. (1994a)
K. Jormakka, How to read an object as art. Nordic J. Arch. Res. (1994b)
K. Jormakka, D. Kuhlmann, O. Schuerer, *Design Methods (Basics)* (Birkhäuser, Basel, 2007)
D. Kuhlmann, K. Jormakka, *Ars Imitatur Naturam*, Thesis, vol. 6 (Bauhaus Universität Weimar Verlag, Weimar, 1997), pp. 6–13
D. Kuhlmann, *Lebendige Architektur* (Bauhaus Universität Weimar Verlag, Weimar, 1998)
Le Corbusier, *L'art decoratif d'aujourd'hui*, éd. by Arthaud (Paris, 1980), p. 176
Leopardi, G. (1966) in *Selected Prose and Poems*, ed. by I. Origo, J. Heath-Stubbs (Oxford University Press, London, 1966)
G. Lynn, *An Advanced Form of Movement*, in *Architectural Design*, vol. 67 (1997), pp. 54–59

G. Lynn, Animate form. *Faultlines*. UCLA Department of Architecture and Urban Design. 1997, vol. 1, No. 2, p. 1ff

S. Mallarmé, in *Selected Poetry and Prose*, ed. by M.A. Caws (New Directions Books, New York, 1982)

B. Mandelbrot, *The Fractal Geometry of Nature* (W.H. Freeman, San Francisco, 1982)

B. Mandelbrot, Fractal as a morphology of the amorphous, in *Fractal Landscape from the real World*, ed. by B. Hirst (Cornerhouse, Manchester, 1994)

G.E. Moore, *Principia Ethica* (Cambridge University Press, Cambridge, 1959)

F. Neumeyer, *The Artless Word. Mies van der Rohe on the Building Art*, trans. M. Jarzombek (MIT Press, Cambridge, 1991)

F. Nietzsche, *The Will to Power* (Vintage Books, New York, 1968)

F. Otto, *Biology and Building 2. Practical Application of Analogous Research*. IL 4, (University of Stuttgart, Germany, 1972)

F. Otto *Pneu and Bone. The Structural System Pneu in Living Nature*. IL 35, (University of Stuttgart, Germany, 1995)

D. Sharp, Organische Architektur — eine Art, Natur zu verstehen und wiederzugeben? *Archithese* **5**, 17 (1988)

R. Shusterman, *Pragmatist Aesthetics. Living Beauty, Rethinking Art* (Blackwell, Oxford, 1992)

P. Sloterdijk, *Sphären. Eine Trilogie* (Suhrkamp Verlag, Frankfurt, 2005)

L.H. Sullivan, *Kindergarten Chats and other Writings* (Dover Publications, New York, 1979)

C. Thacker, *The Wildness Pleases. The Origins of Romanticism* (Croom Helm, London and Canberra; St. Martin's Press, New York, 1983)

D.W. Thompson, *On Growth and Form* (Cambridge University Press, Cambridge, 1942)

Vidler, A, Case per cyborg: protesti domestiche da Salvador Dali a Diller e Scofidio [Homes for cyborgs: domestic prostheses from Salvador Dali to Diller and Scofidio]. *Ottagono* 96, 36–55 (1990)

A. Vidler, *The Architectural Uncanny, Essays in the Modern Unhomely* (MIT Press, Cambridge, 1992)

E.E. Viollet-le-Duc, *Dictionnaire raisonné de l'architecture française du XIe au XVIe siècle* (A. Morel, Paris, 1875–1882)

W.K. Wimsatt, M.C. Beardsley, The intentional fallacy, in *Critical Theory Since Plato*, ed. by H. Adams (Harcourt Brace Jovanovich, New York, 1971)

Chapter 9
Fractal Geometry of Architecture

Fractal Dimension as a Connection Between Fractal Geometry and Architecture

Wolfgang E. Lorenz

Abstract In Fractals smaller parts and the whole are linked together. Fractals are self-similar, as those parts are, at least approximately, scaled-down copies of the rough whole. In architecture, such a concept has also been known for a long time. Not only architects of the twentieth century called for an overall idea that is mirrored in every single detail, but also Gothic cathedrals and Indian temples offer self-similarity. This study mainly focuses upon the question whether this concept of self-similarity makes architecture with fractal properties more diverse and interesting than Euclidean Modern architecture. The first part gives an introduction and explains Fractal properties in various natural and architectural objects, presenting the underlying structure by computer programmed renderings. In this connection, differences between the fractal, architectural concept and true, mathematical Fractals are worked out to become aware of limits. This is the basis for dealing with the problem whether fractal-like architecture, particularly facades, can be measured so that different designs can be compared with each other under the aspect of fractal properties. Finally the usability of the Box-Counting Method, an easy-to-use measurement method of Fractal Dimension is analyzed with regard to architecture.

9.1 Fractal Concepts in Nature and Architecture

9.1.1 From the Language of Fractals to Classification

For a long period of time, nature has been an inspiration for architects, which implies copying natural forms, translating them into floral ornamentation or using underlying structures found in nature for static optimization and many other possibilities

---

W.E. Lorenz (✉)
Digital Architecture and Planning (IEMAR) E259.1, Institute of Architectural Sciences,
Vienna University of Technology, Treitlstrasse 3/1, 1040 Vienna, Austria
e-mail: lorenz@iemar.tuwien.ac.at; www.iemar.tuwien.ac.at

of translation. Fractal Geometry has provided scientists with an improved approach to analyzing and generating natural forms. In 1975, the computer-scientist Benoit Mandelbrot introduced the term Fractal to describe irregular, non-smooth curves, and to distinguish self-similar non-smooth structures from smooth Euclidean ones. When he writes that clouds are not spheres, mountains are not cones or bark is not smooth, this shows very clearly that Euclidean geometry lacks the capability to describe natural objects [1]. With his concept of Fractal Geometry, Mandelbrot has increased and broadened our insight in nature. Over two thousand years, our environment had primarily been described in terms of classical Euclidean Geometry – the geometry of simple shapes – and people had focused on a simplified view of nature. Fractal Geometry, however, offers methods to describe and produce nature-like objects directly, using the underlying structure rather than describing them with simple forms or reducing the overall form by dividing it into more simple, smooth components. The language of Fractals enables us to describe the twisted, rough and irregular surfaces of our environment by a few simple rules without reducing their complexity. The major question in this article is whether the language of Fractals may also be applied to architecture.

The language of Fractals can be illustrated with the help of the Barnsley fern, a computer generated fern named after the American mathematician Michael Barnsley (Fig. 9.1). It is supposed to resemble the Black Spleenwort, Asplenium adiantum-nigrum. At first sight, the shape of the natural fern can only be described precisely by defining each detail in an extremely time-consuming process. This type of description could be compared with a non-compressed computer image, saving information about every single dot in large files. Describing a facade in that way would mean providing information on size and position of every single architectural element, including roof, doors, windows, window-strips and columns down to even small-sized ornaments. Barnsley used the iterated function system (IFS) to generate an image of the fern with only four relatively simple transformation rules. Four different configurations of translating, reducing and rotating an initial rectangle or converting it into a single line – and the right fine-tuning – are the basis for a fern-like image. In that process the overall information is reduced to a few underlying construction rules – the algorithm is then the language that is used to describe Fractals. Even though the complex output may not be an exact copy of its natural

Fig. 9.1 Barnsley fern

counterpart, the character is the same and, by fine-tuning, the image can become a very close approximation. How can we draw an analogy between that approach and methods found in architecture? Some architects used basic ideas and basic motifs as a designing tool. Horizontality for instance can be regarded as the basic idea of Robie House by Frank Lloyd Wright – not only for a first impression but also for a deeper understanding.

Focusing on details of the fern we become aware of its fractal characteristics. No matter which part is analyzed, it looks like the whole or, putting it differently, the same characteristics can be discovered on each level of scale – the basic settings of the configuration are present in each part. This phenomenon is called self-similarity, and one of the main properties of Fractals is as follows: smaller parts of an image are hierarchically linked to the whole. Analyzing nature in terms of Fractals shows that nature, from mountains to coastlines and down to plants, is based on self-similarity – it is self-similarity that makes nature so fascinating from the large to the small. In case of Robie house horizontality is evident in the overall view as well as in the horizontally stretched roofs, window-strips and even in details such as long-stretched bricks. The basic idea of horizontality is the common denominator for all the individual components that form a complex whole. As the example of the fern illustrates, Fractal Geometry can be used to simulate complex natural objects with the help of simple algorithms, even if, in many cases, no clear rule of the object's development can be identified at first sight. That is, the right configuration of translating, reducing and rotating has to be found first. But as soon as the right configuration has been detected, complex natural objects can be described with the help of Fractal Geometry. This may also be true for complex artificial objects.

In architecture Mandelbrot distinguished buildings of the Beaux-Arts, which are close to Fractal Geometry, from buildings by Mies van der Rohe, which he calls scale-bound throwback to Euclid [1]. It seems that Euclidean Geometry of Modern architecture – throughout this study the term of Modern Architecture is used for those styles using the design vocabulary of simplification of form and the elimination of ornament – is mostly reduced to a few elementary objects. Even if every single architectural element of a smooth architectural design has to be defined by its size and position, the resulting data set remains manageable. But, would this mean that the Fractal concept is not valid for classical Modern architecture? At first sight, no connection between different elements can be found that might reduce the data set to a set of a few construction rules or to basic ideas. But on closer observation there are such rules, for instance rules developed out of systems of proportion that are applied to all elements from the overall facade to the very small detail. Fractal-like architecture is characterized by the presence of various details of several different sizes that are linked together. Le Corbusier demonstrates how this connection can even be based on a certain angle only, e.g. at the facade of a villa he designed in 1916 [2]. There a specific angle defines the diagonal of the overall facade whose numerous parallels together with their perpendicular lines determine the dimension of elements of the second order such as doors, windows down to certain details. Le Corbusier created a similar example with House Ozenfant in 1923. In those examples all elements such as windows and individual parts but also the whole are based on a system of proportion. The only difference between these examples and architecture that is regarded as more fractal-like is the limited range of scales.

If architecture is analyzed from the point of view of Fractal Geometry, an index of coherence can be introduced. On the very low end, there is an absolutely smooth plane then, an empty rectangle, which is not fractal but belongs to Euclidean Geometry. The index of coherence is increased if fractal properties such as self-similarity and others, which will be described in the following section, are present. Facades belonging to this section are then called fractal-like, because they are not Fractals in a mathematical sense, but offer fractal properties within a limited range of scales. In case of such facades, single architectural components of different sizes are combined, leading to a consistent overall composition. At the upper end of the index of coherence, we can find architectural examples with many unlinked details, hence confusing patterns. "Unlinked details" means that architectural components of different sizes are not interrelated, for instance by a formal basic idea. That means while approaching such a building the observer is confronted with constantly changing and often confusing new impressions.

The systems of proportion at the facade of House Ozenfant are only applied within a restricted range of scales and, in addition to that, the facade looks rather smooth. Those are reasons that put the facade of House Ozenfant on the lower end of the index of coherence, while Robie House can be placed in the middle because of its self-similar characteristics.

9.2 Fractals: A Definition from a Mathematical and an Architectural Point of View

Fractals can be explained by their properties. These include roughness, self-similarity, development through iterations, infinite complexity, dependence on starting conditions, they are common to nature and their Hausdorff dimension exceeds their topological dimension. A simplified definition of the Hausdorff dimension can be illustrated by covering a set of points of finite expansion in a three dimensional space with a minimal number of balls $N(R)$ of radius (R). Decreasing the radius will enlarge the number of spheres. The Hausdorff dimension (d_H) is then defined by

$$d_H = -\lim_{R \to 0} \frac{\log(N)}{\log(R)}.$$

9.2.1 Roughness and Length Measurement

Roughness can be described very well if we look at the coastline of Norway, where Fjords with their sub-bays, inlets, cliffs and rocks lead to a very fractional border in contrast to a circle offering a smooth border. This also becomes evident in connection with length measurement. Benoit Mandelbrot introduced Fractal geometry with the question: "How long is the coastline of Britain" [1]. This is not a trivial question. For length measurement, different maps of various scales can be used, which represent different distances between the observer and the coastline. A large scale may correspond to the view out of an aeroplane hundreds of meters above ground, while a small scale may correspond to the impression an observer gets when walking along the coastline on foot. If we keep in mind the example of the Fjords, it is obvious that each time a smaller part of the coastline, represented on a map of smaller scale, is analyzed, new sub-bays will become visible, which could not be identified in the previous large-scale map. The length measurement of the coastline on a map using a larger scale presents a rough image of the real coastline as it only includes the larger bays. If the same section of the coastline is measured on a small-scale map offering a more detailed version of reality, the larger bays mentioned before are indented by smaller bays so that the measured length increases. In other words, length depends on the details presented on the map used, or on the measuring devices in the real world. In contrast to the Fjords, if the borderline of a circle is analyzed in the same way, the length measurement will follow a different behaviour. If we use smaller and smaller measuring devices, length will tend to a limiting value very quickly (approximation through polygon with "N" edges). This is because no additional details will be presented while zooming in. The difference between Fractals and Euclidean objects (e.g. squares and circles) then lies in the fact that length measurement fails for the first ones.

What does length increase look like in a mathematical fractal such as the Koch curve, a famous representative of a classical Fractal? The construction rule of the Koch curve starts with a straight line of length one, which is called the initiator. This line is then divided into three equal parts (Fig. 9.2). The middle part is replaced by an equilateral triangle, whose base line is removed. This adjustment of four smaller lines is called the generator of the Koch curve. As the resulting structure consists of four scaled-down copies (four lines) of 1/3rd of the initiator (one line of the initial size one), the overall size increases by 4/3rd. In the next step, each of the four new lines is again replaced by the generator, resulting in 16 scaled-down copies. The scaling factor in relation to the initial line is now 1/9th. That means the length increase is 4/3rd by 4/3rd or in other words 16/9th. From this it was not only followed that the length increases by 4/3rd from one iteration – that is a step of replacement – to the next, but also that the offsets of the curve increase – a higher degree of roughness can be identified. For the third iteration length increase – in relation to the initiator – is 4/3rd by 4/3rd by 4/3rd or 4/3 above 3. Expressed in a more general way this leads to the following equation

$$L_{(i)} = (N_{(1)} \times s_{(1)})^i$$

with (i) indicating the number of iterations, ($L_{(i)}$) the length of the curve after (i) iterations, ($N_{(1)}$) the number of single lines of the generator and ($s_{(1)}$) the reduction factor of the generator, being one third for the Koch curve. From one iteration to the

Fig. 9.2 Koch curve

next, the curve gets longer and longer, hence more and more twisted, which means it is getting rougher from iteration to iteration. Because mathematical Fractals are results of infinite iterations, the final length is infinite.

9.2.2 Scale Range and Distance

As the above example of coastlines shows, scale range and distance determine the degree of roughness and detail the viewer is able to see. These factors also influence a viewer's perception of a building. From far away, the viewer will only perceive very large components, the silhouette and significant edges. This view corresponds to an elevation of large scale that only includes a few details. Approaching the building, the observer's attention begins to focus upon the sequence of base, middle or roof part. Then windows and doors or the rhythm of columns are the most prominent parts. Zooming to one of the next levels it turns out that (e.g.) windows consist of smaller details such as window frames and window handles, but even walls offer smaller parts such as face bricks or tiles. Thinking about grain of wood, the cascade of details will come down to the material itself. For Salingaros, the size of the smallest detail has to correspond to the smallest perceivable scale [3]. This clearly shows that facades are in general not smooth Euclidean two-dimensional planes, which is per se valid for Modern architecture as well as organic architecture or any other style. Elevations as two-dimensional translations of facades are rather something between lines, defined by edges, and the plane, defining the surface. Before architecture can be analyzed and buildings compared with each other from the point of view of coherence between elements of different sizes, or more precisely with regard to self-similarity, it is important to define the range of scale first – it has to be standardized. The range correlates with the distance of the observer while approaching the building and the smallest possible detail that can be perceived from certain distances.

9.2.3 Self-Similarity: An Important Attribute of Fractals

Self-similarity is an important property of Fractals. In mathematical terms, two objects are similar if their corresponding angles are identical and their corresponding sides are in proportion regardless of their size. An object is called self-similar if one or more of its parts look like the whole. The parts can be exactly similar or approximately similar. The Koch curve is exactly self-similar because each part is an exact, scaled-down copy of the whole. The final Koch curve is the result of applying similarity transformations (scaling, translation, rotation) to the initial object, which modifies proportion by the same factor. The resulting scaled-down copy may be rotated or transformed while the shape remains the same. If the pieces of the object are scaled down by different amounts in different directions, the fractal is then called self-affine instead of self-similar. When simulating nature the factor of chance has to be added. The new structures are called statistically self-similar, that is whenever small copies, looking like the whole, have variations [4]. Parts show the same statistical properties at many different levels of scale. This is important for describing natural objects, but also architecture. When we, for instance, examine certain parts of coastlines, they do not just represent scaled-down, transformed and rotated copies of the whole, but show a similar character and degree of irregularity. Cliffy coastlines offer the same strong irregularity from the very large to the very small level of scale, similar to gently twisted coastlines offering the same softness all over as well.

Geometric shapes with fractal properties had already been known long before the term Fractals was introduced and long before all the facts about Fractals were combined to form a theory. The Koch curve for instance – a continuous curve without tangents – was already presented in 1904. In connection with length increase, it was indicated that the number of components (single lines) of smaller scales increases from one iteration to next. In architecture, fractal properties had also been used consciously but also unconsciously, which is true for Gothic cathedrals as well as for Robie house by Frank Lloyd Wright [16, 17]. Carl Bovill already drew an analogy between writings of Frank Lloyd Wright and the fractal concept [5]. According to Frank Lloyd Wright, nature should be a source of inspiration. Carl Bovill demonstrates that for Frank Lloyd Wright the underlying organizing structure of nature – which is finally the fractal concept – is of particular interest. The fundamental idea of bringing the characteristics of a horizontal environment, the Prairie, into the building by Frank Lloyd Wright, Robie House, is implemented on many levels of scale, from the wide overhanging roof and stretched storeys over window strips and horizontal parts of walls to the design of the horizontal and straight joints [15]. The number of components that are evident on a specific level of scale increases from the overall view over the level of scale that includes windows down to ranges of scale on which details of bricks and material become visible. Analogously, a cauliflower consists of a smaller number of similar scaled-down components that again consist of many scaled-down components. In this sense examples such as Robie House can be called fractal-like architecture. Architecture is called fractal-like if the whole and all other formal elements are derived from one basic idea and, by that, a simple, specific form characterizes the expression of the building. Parts are then reflections of the whole, the formal elements are held together in scale and character [5]. Eero Saarinen also pointed out that a building should follow the strong, simple concept of the whole – each part has to be an active component of a certain overall theme. This is valid irrespective of the point of time when decisions to follow such an overall theme are taken: Right from the start with regard to ground plan or construction-system or at a later stage, when detailed elements such as colours inside the house or even door-handles are in the architect's focus of interest [6].

9.2.4 Architectural Examples

There are different approaches to self-similarity in architecture: Self-similarity in architecture may arise from a basic proportional system, from a basic form or from an overall idea. To give another example, Rietveld-Schröder house in Utrecht (Netherlands, 1924) offers coherence from the large elements to the small ones [17]. The architect Gerrit Rietveld translated the use of form of the Stijl, with basic forms and basic colours, into architecture. Those basic forms are taken from Euclidean Geometry and include straight lines, planes and slabs that are detached from each other and seem to glide past – the floating space playing an important role [7]. The building is characterised by large openings, horizontal and vertical elements and an intersection between inside and outside elements. The cascade of architectural components starts with basic spatial and constructive structure, formed by few main elements of large white slabs (windows and doors are cut out). Then, on a smaller level of scale smaller slabs for balconies and canopies follow, while linear elements such as rainwater pipes add vertical and horizontal accents. Inside, an analogy can be drawn between the basic elements and window-frames, sliding and revolving panels and their linear hanging. Coming even closer, the basic elements are repeated in the design of the furniture, from chair to modular cupboard. The translation of basic elements into smaller levels of scale connects the furniture to the architecture around. This example illustrates that even though the number of iterations is limited, the basic idea of separating components by means of form and colour and the composition of planes and linear accents is nevertheless evident within all levels of scale.

There is one more example we should deal with: in 1956, Bruce Goff constructed the Joe Price studio in Bartlesville, Oklahoma. The basic idea of construction was focused on 60° angles and their multiplication or subdivision to hexagons and triangles [8]. Once again self-similarity occurs on the basis of a simple component, a certain angle. Variations of the basic element can be found from the large elements, including the roof, with the basic shapes forming a twisted outline, down to the small size of decorative elements. Inside in turn, the ceiling, the walls, the hexagonal sitting hole, which is a hollow in the floor, and certain details are based on triangular shapes. The individual components of different sizes are variations of the basic theme and therefore offer coherence within all levels of scale. Such variation arises from the material used and from purpose such as deflecting sound in the music room.

The work of Antoni Gaudí offers fractal-like architecture as well [16]. Although buildings by Antoni Gaudí appear very complex, they are nevertheless coherent. This is not a surprise because Gaudí pointed out the importance of dealing with details to produce a complete work of art. As an example the use of curved construction stones and nature-like, organic forms can be found all over Sagrada

Familia from outside to inside from large elements to small details. Gaudí was interested in the forces of nature that act behind the surfaces rather than in the shape, hence the surface itself [9]. Inspired by nature, Antonio Gaudí introduced catenary shapes as idealized forms of arches [10]. He developed catenarian-models and cable-models with sandbags illustrating the reverted interrelation of forces for columns and pillars. The result was then an upside down model of the structure in the building. Gaudí found out that the use of parabolic arches and inclined buttresses could withstand the forces involved. On smaller levels of scale, hyperbolic paraboloids can be detected in the vaults but also at the base of columns [10]. Antoni Gaudí used cone, cylinder, simple hyperboloids, hyperbolic paraboloids to design non-smooth architectural components. The analysis of a small part of the building will make the observer think about the whole, and the building is coherent in form and character.

Self-similarity alone, however, does not define a Fractal. Considering that a line between two points can also be divided into smaller parts that are scaled-down copies of the whole, the line nevertheless does not increase in length from one iteration to the next, and there are no additional details offered either when zooming in. Or, zooming in on a circle will show a more and more straightened part of the circumference – apart from irregularities arising from the drawing itself – but does not offer additional details. The latter two examples belong to Euclidean Geometry because no further details arise when zooming in.

9.2.5 Developed Through Iteration

Generalized for a first approach, Gothic cathedrals can be described by verticality and light-flooded interior rooms, which was made possible by pointed arch, flying buttress and ribbed vault. They are an expression of unity between single components, inside and outside, and the whole: In numerous functional and decorative elements such as windows, portals, baldachins, pinnacles, attics, Gothic gables and tabernacles, references to the whole can be identified. Examples such as the Gothic windows of the Southwest tower of the cathedral of Cologne or of the Angel Choir of the Cathedral in Lincoln definitely provide excellent examples to describe how Fractals can be constructed. The overall shape, a pointed arch, defines the initiator, which is then replaced by the generator, a pointed arch of the same size vaulting two smaller ones of half the size. In the next step, each pointed arch is again replaced by the generator and so forth. Figure 9.3a shows a series with the initiator that is the starting object, the generator that is the replacing rule and some iterations. Although the replacing rule for the Gothic window may be applied infinite times in theory, in the real world there are some restrictions arising from the material used and from the usability as a window. The built version tries to find a harmonious balance between a slight big opening and construction rules and therefore already ends after a few levels of replacement or iterations, respectively. Furthermore, the algorithm may be adopted with regard to the reduction of the diameter of the shafts.

Such replacement procedures can also be identified as underlying construction elements of Indian temples, such as the Sikbam of a Jaina temple on Mount Girnaz [11]. The second image in Fig. 9.3b shows an output of a construction rule that is derived from this example. The basic curved shape of the whole temple, rising out of a square, has undergone a transformation in size and position, bringing up four additional similar shapes at the sides of the main part. Each of these four new elements is then transformed in the same way, resulting in a complex whole, where all parts are formally linked with the whole. Figure 9.3b shows the initiator and the fourth iteration, each of three different configurations of such an insertion rule. The basic transformation rule, scaling down the initial shape and positioning it at each side, is adjusted in the way that different reduction factors and movements are used. It is also conceivable that the reduction factor varies within the same iteration by the factor of chance. Using different basic shapes (initiator) and adjustments, the basic rule may then lead to high rising as well as vaulted structures (Fig. 9.4a, b).

The output of such computer-generated self-similar structures cannot be used for architectural design without changing or adapting it with regard to the limits imposed by the material used, construction, function and environmental influences. Such computer outputs are only useful to show how self-similar patterns in connection with architecture can be generated in a simplified way. If facade generating programs are developed to compute facades with visual depth, the question of fabrication and usability has to be considered. Such facade generating programs might start with a simple box, dividing it in (n) by (m) boxes, which are removed or added in certain ways. Single boxes vary in their expansion to the front, so that variation in depth is achieved. Then the new boxes are manipulated in the same way, using the same parameters or slightly adopted ones with, e.g. random

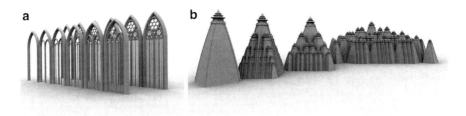

Fig. 9.3 Various examples of geometric output – (**a**) Gothic window algorithm, (**b**) temple algorithm

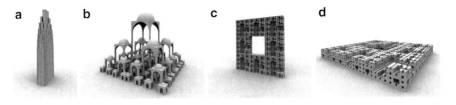

Fig. 9.4 Different insertion rules – (**a**) high rising, (**b**) vaulted, (**c**, **d**) window element

factors for expansion. Figures 9.4c, d are based on such a rule and offer self-similar patterns after a few iterations. Windows are either included in such algorithms by cutting out middle parts or such outputs are simply arranged around existing cuts. In the first case, one large cut in the facade is surrounded by smaller ones, surrounded by even smaller cuts and so on.

9.2.6 Differences Between Architectural and Mathematical Fractals

The main difference between nature or fractal-like architecture and mathematical Fractals is the limited number of iterations. Although mathematical Fractals are theoretical constructions offering infinitely small parts, self-similarity of nature and architecture only exists for a limited range of scale. It could be said that nature and architecture only uses a limited number of iterations. This will be illustrated by comparing the Sierpinski triangle, a mathematical fractal, with Castel del Monte, situated in the Apulia region, by the Holy Roman Emperor Frederick II. The Sierpinski triangle can be simulated by starting with an equilateral triangle as the initiator. This initiator is then replaced by a generator consisting of three equilateral triangles, each scaled-down copies of the initiator by the factor of 0.5. All corners of the initiator act as fixed points for one of those three new triangles when they are scaled down. In other words an equilateral triangle is cut out in the middle of the initiator where the corners of this cut out triangle are situated on the middle points of the sides of the initiator. For the next iterations, each triangle is subjected to the same procedure, leading to nine triangles. In theory, this is repeated infinite times leading to a real Fractal, where even the smallest part is a scaled-down copy of the whole. In case of Castel del Monte, which offers the beginning of a replacing algorithm, the difference between mathematics and architecture becomes evident. The basic octagonal shape is complemented by further octagons at its corners. For the next iteration, each of these octagons would be complemented by further octagons at their corners and so on. But the "real" Castel del Monte only offers the first iteration. Nevertheless, the circumference of Castel del Monte increases from the initiator to the first iteration and offers a rough outline. Together with the rough stonewall this distinguishes Castel del Monte from smooth surfaces with a lower index of coherence.

9.2.7 Fractals as a Design Aid

The competition contribution for Cardiff opera house in Wales by Greg Lynn in 1994 mandates a new concept for waterfront urban space that is nonetheless in conformity with the history of the site and Cardiff's waterfront [12]. The analysis of the coastline for self-similar structures from the large down to the small level of scale (water's edge being captured by land) together with the oval form of the basin was translated into a rule with the starting body being replaced by three parts of different sizes and orientation. Some iterations of this Fractal basic shape in combination with the basin defined the character of the opera house next to it – the opera house is then a continuation of Cardiff's waterfront. The output was adapted with regard to function, construction and form. Volumes were rearranged because of requirements of the foyer, auditorium and acoustic properties, stages, studios, offices and other purposes of the opera house.

In general, fractals can also be used as a basic design for breaking open an otherwise straight line. For example, the characteristics of a natural coastline can be simulated by a fractal, if the right configuration or the right insertion rules are found. The twisting of the Fractal then leads to a coastline section, which is based on a more environmentally appropriate scale. What can be learnt from such attempts – forming buildings and border-lines? In general, using such Fractal-based designs means that the architect looks for a rule coming close to the one that is inherent in the environment on a larger scale to continue the cascade of similar character down to the size of harbors or buildings and even to human level connecting smaller scales to the whole. Man-made interaction will then continue the natural characteristics of the environment and in turn will not restrict the range of scale. The basic Fractal nevertheless has to be adapted with regard to usability concerning function and costs, but also with regard to the fact that the result may again consist of straight parts though on an even smaller scale.

9.2.8 Fractals Are Common to Nature

Dealing with Fractal Geometry also means focusing on nature: Clouds, bark and trees are not smooth but rough, and snowflakes or the distribution of stars offer self-similarity. Different fractal methods, which take advantage of self-similarity, underline the connection between nature and fractals [20]. The right algorithms produce models of plants, mountains, crystals and entire natural scenes. Self-similarity may in turn be the reason why nature seems so fascinating from the large elements to the small ones and why forms of nature seem better balanced to us than Euclidean smooth shapes. If so, this explains why Gothic cathedrals, rural houses and organic buildings, which rather contain self-similarity, are so fascinating to many observers. Frequently those buildings offer details, which are prominent on different levels of scale and are consistent with each other and the whole. Consequently, while we are approaching the building, new details of smaller size get into the focus of our attention reminding us of the whole. Because of variation of components, which differentiates self-similarity from self-sameness, they also remain diversified. Assuming that those parts are exact, scaled-down copies of the whole, this would mean the observer can judge from looking at the whole what the detail will exactly be like. Because of variation he can envisage what comes next but may be slightly surprised and confirmed at the same time. Mandelbrot believes that fractal art is more acceptable because it imitates nature to make the observer guess its rules and is therefore more familiar to us [1].

9.2.9 The Factor Chance

Nature-like images produced by strict self-similar construction rules are often too clinically "perfect" to imitate their originals to their last consequence. Local influences such as temperature, wind, waves or nutritive substances, which deform the objects, are missing. Effects of such influences can be imitated by the factor of chance. This can be illustrated by means of the tree algorithm (Fig. 9.5a). Starting from a stem of certain diameter and length, three or any other number of scaled down copies of it are moved to the end of the stem and rotated in different directions. Each of these branches is again copied, scaled down, moved to the end and rotated in different directions. This rule is applied to a couple of iterations. If the rotation is chosen randomly between certain limits as well as the length of each branch, the resulting image resembles nature even more closely. The basic shape is a smooth Euclidean cylinder, but the object itself becomes rougher from one iteration to the next, while smaller pieces are added. On the one hand, the resulting tree can be described by defining length, position and direction of each single branch, which produces large data quantities. On the other hand, the resulting tree can also be described just by its insertion rule.

With the factor of chance, even the Koch curve can be turned into a natural-looking coastline. Figure 9.5b shows three such Koch curves added to one Koch Island. In the basic rule, the middle vertex of the generators of each Koch curve looks to the outside and so does every replacing step. To get a more nature-like model, the middle vertex is also allowed to look to the inside. The choice whether it is pointed to the outside or inside is chosen randomly. After some iterations, the resulting curve is not as clinical as the origin without the factor of chance, but nevertheless offers the same characteristics and length increase. Both examples, the tree

Fig. 9.5 (a) Simulation of a tree, (b) Koch Island with factor of chance, (c) simulation of a mountain

algorithm and the coastline, indicate the importance of variation for natural-looking images. This has been taken into account for certain Fractal methods producing images of plants, trees, mountains and even planets. The rule used for such a method can be very simple, while the output will look extremely complex. This can be illustrated with a program imitating mountains, which can be implemented into a CAD package quite easily. The starting image to generate such a mountain is a triangle. Then the midpoints of each of the three sides of the triangle are marked. These points are then moved up or down by chance and by a certain factor. The higher the factor, the rougher the resulting mountain is. In the next step, each point is connected to a new triangle with its neighbouring points. These triangles are then subjected to the same rule as before (Fig. 9.5c). Parts are not identical with the whole but the overall character remains the same from one zoom level to the next – analyzing small parts will offer similar structures as the whole. From this follows that, the other way round, underlying construction rules of natural objects are difficult to identify because of variation, which have been here simulated by the factor of chance.

9.3 From Simulation to Measurement

Fractal methods generating architectural structures, as they have been introduced with the Gothic window, the Indian temple and the facade designer, are only a first approach to Fractal architecture. All those attempts have in common that their algorithm is very simple, just demonstrating basic rules of existing buildings, simulating them with the help of a computer and producing outputs of different shapes. This first attempt can be made more sophisticated with the factor chance to develop symmetry breaking designs such as those that may result from environmental restrictions or adjustments because of internal function. Including parameters that simulate the influence of daylight and shadow, compactness, functional fitness or resources directly will increase computing time. With the help of a computer, a large number of different alternatives can be generated from which the architect can finally make his choice. This choice can then be modified and developed further as in the case of the competition contribution for Cardiff opera in England by Greg Lynn.

9.3.1 Curdling

Several examples of fractal-like architecture from Gothic buildings to Robie house and simplified architectural simulations have not only shown that, similar to nature, architecture possesses fractal properties up to a certain degree, but also that those properties are difficult to describe because of modification by certain influences. Consequently, for comparing works of architecture with each other – with regard to the degree they are fractal-like – a consistent measurement method has to be developed. If self-similarity is present in architecture, then this is also expressed by a similar distribution of architectural elements from one level of scale to the

9 Fractal Geometry of Architecture

Fig. 9.6 Curdling

next – each level of scale has its elements of specific size that again contains smaller details of similar distribution. Such connections between different levels of scale can be demonstrated with the help of the so-called curdling, introduced by Benoit Mandelbrot to demonstrate the process that produces a disconnected set of points with nevertheless clustered characteristics [5]. The process is called curdling since originally uniform mass distribution clogs together forming many small regions of high density [4]. There, structures are generated by the factor chance where zooming in and analyzing parts will offer similar characteristics like the whole. Finally, such connection between many levels of scale can be given by a set of values, characterizing the cluster. The measurement method of these values will now be introduced with curdling.

"Curdling" was the name Mandelbrot coined for a procedure that produces a random fractal dust in two dimensions [1]. With curdling, the starting object, a simple square, is divided into a grid. In our example, the grid consists of three by three cells. Tossing a coin determines whether a cell is deleted or again divided into a three by three grid. The coin can also be replaced by a probability factor. High probabilities lead to a higher chance for cells to remain and consequently to a higher 'density'. Figure 9.6 shows some iterations for different probabilities.

Considering the probability of one out of nine, this means that mathematically one cell will remain after the first iteration. That also means that eight out of nine cells are deleted. Because of the element of chance, in practice none may be chosen as well, which stops the algorithm, but also more than one is possible. What happens to the number of cells when using the probability two out of nine? Mathematically, after the first iteration only two of nine cells remain and seven are deleted. For the second iteration, both remaining cells are again divided into a three by three grid. Again in each grid, two cells remain and the others are deleted. This increases the total number of remaining cells up to two by two, hence four. In the next iteration, each of these four cells is again divided into a three by three grid. Again two cells of each grid remain, increasing the total number of remaining cells to two by two by two, hence two above three. Simplified, this leads to the equation

$$N_{(i)} = (N_{(1)})^i$$

giving the connection between increasing numbers of remaining cells and iteration. There $(N_{(i)})$ is the total number of remaining cells after (i) iterations and $(N_{(1)})$ the number of remaining cells of the generator. For the probability three out of nine, the number of remaining boxes after three iterations is then three above three, hence $N_{(3)} = 27$. For the same iteration but for the probability eight out of nine it is $N_{(3)} = 512$.

Simulations use a random generator of a certain probability that determines for each cell whether it remains or not. Therefore, different simulations of the same probability may lead to different numbers of remaining cells. The two examples on the right in Fig. 9.6 show two different results for the probability of two out of nine, the middle two for the probability six out of nine and the two on the left for the probability eight out of nine. These examples indicate that the remaining cells after (i) iterations vary for one and the same probability. For simulations, the simple connection between the number of remaining cells and iteration of the previous equation is then no more valid.

Considering that the grid represents a building site, the same algorithm may simulate distribution of buildings. This time the grid does not have to be regular and the algorithm will be stopped at a proper size, that of buildings. Then cells of the last iteration are moved from the border defining streets in between. Once again this is just a formal simulation using one basic rule only, but it will produce choices of different distributions as discussion bases. More practicable models can be developed, enlarging the basic rule by additional parameters accounting for different influences.

9.3.2 Fractal Dimension

For indicating consistency between different iterations, we have to look for a comparison between the increase in number of remaining cells and the reduction factor. With the Koch curve, the number of single lines ($N_{(i)}$) and the reduction factor ($s_{(i)}$) are both increasing/decreasing by the same index: number of parts after the first iteration ($N_{(1)}$) = 4, the reduction factor of these parts compared with the size of the initiator ($s_{(1)}$) = (1/3); number of parts after the second iteration ($N_{(2)}$) = 4^2, ($s_{(2)}$) = $(1/3)^2$; ($N_{(3)}$) = 4^3, ($s_{(3)}$) = $(1/3)^3$. From this it was derived that there exists a connection between the number of single lines after (i) iterations ($N_{(i)}$) and the reduction factor ($s_{(i)}$). This connection is shown in the equation

$$N_{(i)} = \left(\frac{1}{s_{(i)}}\right)^{D_s}$$

introducing index (D_s), which is the self-similar dimension. Modified, this leads to

$$D_S = \frac{\log(N_{(i)})}{\log\left(\frac{1}{s_{(i)}}\right)},$$

which equals (D_s) 1.26 for the Koch curve with four lines of one third each after the first iteration (i) = 1.

9 Fractal Geometry of Architecture

With the theoretical output of curdling, self-similar dimension can be calculated in the same way. Using the example from above – tiling each cell into three by three smaller cells – with a probability of two thirds, this means that mathematically six cells will remain ($N_{(1)}$) after the first iteration. The reduction factor of the grid-size is one third ($s_{(1)} = 1/3$). Inserting these values into the last equation, the self-similar dimension (D_s) equals 1.631. With the probability of one ninth, mathematically only one cell remains after the first iteration while the reduction factor is again one third. This leads to the value zero, equal to the topological dimension given for a dot. At the other extreme with the probability of nine ninth all nine cells remain, resulting in the value two, which equals the topological dimension for a plane.

What is of interest next is how the value (we will call it the index of coherence), defining the connection between the number of single elements and the reduction factor, behaves throughout the range of scales. With curdling, this means to compare the increase in the number of remaining cells with the decrease of the reduction factor. Using once more the example from above with a grid of three by three cells, mathematically the number of remaining cells after the first iteration is six for the probability of two thirds and the reduction factor is one third. Then the second iteration, using equation $N_{(i)} = (N_{(1)})^i$, increases the number of cells to 36 with the reduction factor decreasing to one ninth. Between two iterations, the differences of both, number of cells as well as reduction factor, can be examined. Consequently, the equation for calculating D_s is rearranged, leading to the equation

$$N_{(i)} - N_{(i-1)} = \left(\frac{1}{S_{(i)}}\right)^{D_B} - \left(\frac{1}{S_{(i-1)}}\right)^{D_B}.$$

The difference of cells is compared with the difference of the inverse reduction factors above the value of index, this time called (D_B) with regard to analyzing boxes (we called them cells). This equation is rearranged to

$$D_B = \frac{\log(N_{(i)}) - \log(N_{(i-1)})}{\log\left(\frac{1}{S_{(i)}}\right) - \log\left(\frac{1}{S_{(i-1)}}\right)}.$$

Inserting the mathematically calculated values of our example with a probability of 2/3rd – that is for the number of remaining cells $N_{(1)} = 6$ with the reduction factor $s_{(1)} = 1/3$ for the first iteration and $N_{(2)} = 36$ with $s_{(2)} = 1/9$ for the second iteration – (D_B) again equals 1.631. Analyzing the third and fourth example from Fig. 9.6, both using a random calculator with the probability of 2/3rd, the results change. While the reduction factor remains the same, the number of remaining cells changes. That is $N_{(1)} = 7$ for the first example after the first iteration and $N_{(2)} = 47$ after the second iteration. The calculated value D_B between first and second iteration then equals 1.73. D_B of the second example equals 1.75 with $N_{(1)} = 7$ and $N_{(2)} = 48$.

So both values are slightly higher than the result inserting the mathematically calculated values. Using the last equation, a set of index values (D_B) can be given for a certain number of iterations. Each iteration is compared with the next one, which leads to a certain (D_B) from the first to the second iteration, to another from the second to the third and so forth. Because of the consistency between iterations, self-similar structures will then offer similar values throughout the range of scales.

9.3.3 Perception and Distance

Comparing buildings with each other means that their appearance has to be standardized first, including influences of color, shadow but also depth of details – equals the number of iterations or levels of scale. For a first approach, just black and white elevations will be analyzed, which means removing all influences other than depth of details. It is the set of index values (D_B) for a certain range of scale that is of interest. Because two-dimensional plans will be analyzed, a first step of standardization is to define the translation of the original building into the elevation. There the smallest detail presented in the elevation depends on the distance of the viewing point from the building in reality and on the human eye. This derives from the fact that the smallest possible detail depends on the reading field, which is inside a cone of $0°1'$ [13]. With the aid of trigonometric function, the relationship between detail and distance can be given by

$$\text{Minimum Size of Detail} > \text{Distance to Detail} \times \text{tangents}\,(0°1')\,.$$

This means that for a given distance of 10 m, the size of the smallest perceivable part is approximately 3 mm. Since the smallest perceivable detail depends on the distance between the observer and the building, consequently the distance defines what should be presented in the elevation. From the smallest detail on, all architectural elements of larger size have to be included and translated to the elevation as significant edges. Only then coherence between these levels of scale or sets of different sizes of architectural elements can be analyzed, but also only then buildings can be compared with each other by their elevations, standardized by the observer's distance. The distance to the building, in turn, should be chosen in the way that its whole extent can be perceived, mainly its vertical extent. Maertens gives an indication, where a distance equal to the relevant height of a building is appropriate to view details of the object [13]. Then the uppermost part is within an angle of 45° above horizon. A distance of double height equals 27° and there the whole building can be viewed for itself. Finally at a distance of the observer of three times the height, the building will become one with its environment. From these angles, the distance may then be derived.

9.4 Fractal Dimension and Architecture

Fractal curves are the result of insertion rules after infinite iterations. The Peano curve, a representative of a classical fractal, is an endless, twisted curve between two points that does not exceed a certain space. As it can be separated by removing just one point from the set, it is said to have a topological dimension of one. But on closer observation, it passes through the two-dimensional plane completely – it offers area-filling property. The Koch curve again does not fill the two-dimensional plane but is also endlessly twisted. Since length is infinite, a point on such curves cannot be defined by only giving the distance. But also a mountain, being rough, does not fill the three-dimensional space completely. Such structures can then be characterised in a better way by their Fractal Dimension. The Fractal Dimension expresses how fast a fractal curve tends to infinity from one iteration to the next or how completely a fractal appears to fill space.

Facades are rough surfaces that consist of cuts, different architectural elements of different sizes and details. Hence, they are no flat smooth two-dimensional planes. Likewise, their expressions on paper, elevations are more than a one-dimensional line that defines the silhouette, but they also do not fill the plane where they are situated in completely. They are rough structures and Fractal Dimension is then the expression of the degree of this roughness, which means how much texture an object has [5]. With fractal-like architecture, self-similarity cannot simply be identified by rescaling parts transforming them into the whole again. Then Fractal Dimension is an adequate possibility to describe such structures, where coherence of roughness can be analyzed by calculating a set of index values (D_B) for a certain range of scale.

9.4.1 Fractal Dimension and Approaching a Building

Expressed in a different way, Fractal Dimension is the degree of mixture of order and surprise. For objects that are visualized on paper, Fractal Dimension can be measured by the Box-Counting Method. This measurement method was first applied to architecture by Carl Bovill [5, 19]. First, a grid is put over the object to be measured. The grid-size is defined by the number of boxes across the bottom row of the grid, that is the numbers of boxes in x-direction ($1/s$). Its inverse value then defines the scale of the grid (s). Then those boxes, which cover relevant parts of the elevation, are counted. Relevant parts are the outline, the roof, windows, doors, walls, but also certain details. The depth of details of the analyzed elevation depends on the scale of plan. For the first grid-size ($1/s_1$), the number of boxes that contain relevant parts is defined as (N_1). For measurement, the grid-size is then reduced to ($1/s_2$), and the number of boxes that contain relevant parts is counted again (N_2). Finally, the Box-Counting Dimension between two scales of grid-size is calculated by the relationship between the difference of the logarithms of the number of boxes that contain relevant parts and the difference of the logarithms of grid-size as given in the equation

$$D_{B(1-2)} = \frac{\log(N_{(S2)}) - \log(N_{(S1)})}{\log\left(\frac{1}{S_2}\right) - \log\left(\frac{1}{S_1}\right)}.$$

The Box-Counting Dimension D_B is another special application of Mandelbrot's Fractal Dimension [4, 14]. The Box-Counting Method compares the roughness – represented by lines – between different grid-sizes and thus allows measuring the complexity of a structure across certain sizes of details. Equivalence exists between the scale of the elevation, the scale of the grid and coming closer to a building. In the first case, a large scale of the elevation only gives an impression of the building. In this case, larger grid-sizes are used for measurement. Then reducing the scale of elevation, which means including smaller details, will allow us to identify more and more details. This asks for smaller grid-sizes. The same is true when approaching the building in reality. On the level of scale of far distance, smaller details are faded out, because they cannot be perceived. Consequently, they have to be excluded from the elevation for measuring the Box-Counting Dimension. Then from a shorter distance, bigger architectural elements such as windows and doors are perceived that could not be distinguished before, followed by window-frames and door-handles. For measuring the Box-Counting Dimension for this distance, the elevation has to include these components. If the building follows the Fractal concept, the kind of roughness nevertheless remains the same for all steps. In the logic of Fractal Geometry, they are linked together by the depth of similar roughness.

9.4.2 Results of Measurement

The behaviour of the relationship between grid-size and number of boxes that contain relevant parts is analyzed in a double logarithmic graph, where the slope of the replacing line defines the Box-Counting Dimension for a certain range of grid-sizes. The result for many different measurements of Robie House by Frank Lloyd Wright remains between 1.6 and 1.65, taking different elevations with different detail-richness into account: from an overview to plans of smaller scale, including details such as stained glass and brick. That means different elevations were used for measurement to include different distances. For the first distance only main edges such as outline, windows and doors were included. Then approaching the building, hence using smaller boxes, sections of the elevation were analysed, including stained glass and bricks. For all measurements, it is valid that the result depends on what is included in the elevation and how it is presented.

The double logarithmic graph illustrates that certain measuring points of Robie House are very close to their replacing lines. From this follows that, although Box-Counting Dimensions between two single grid-sizes may vary when approaching the building, the set of measurement points in the double-logarithmic graph is nevertheless stable over a large range. The slope of the replacing line then gives quite a significant value (D_B) for this range. Comparing these results with the Koch curve, whose Self-similar Dimension is known, it can be indicated that measurements for the Koch curve are even more stable. Nevertheless, while comparing different buildings with each other certain influences have to be dealt with [14, 18]. Some derive from the measurement method itself – that is dependence on starting position or overall grid-size – and others from range dependence of architectural elements. Architectural elements only emerge locally, which means they have a specific range of distance of the observer in which they are significantly present. In general, to minimize local influences a whole range of grid-sizes is analyzed rather than only two single levels. The standard deviation of the graph then gives the degree of coherence. For Robie House, it turned out that the measurement points in the double logarithmic graph are quite stable for a broad range hence indicating coherence for a broad range.

9.5 Conclusions and Outlook

Basically the fractal concept of architecture means that details of different sizes are kept together by a central rule or idea, respectively – avoiding monotony by using variation. In architecture, this concept is the reason why Gothic cathedrals and examples of the so-called organic architecture are so interesting and diversified. Modern architecture may also offer fractal properties but not for a broad range of scale. For measuring the presence and coherence of architectural elements across many levels of scale, the Box-Counting Method turned out to provide a first verifiable measurement method. The double logarithmic graph – grid-size vs. number of boxes covered – gives a first indication for similar density across certain scales. Although the resulting graph does not tell us anything about the quality of a building or about its form, it provides a first impression of the coherence between levels of scale as it is true for Robie House by Frank Lloyd Wright.

Future focus lies on further Box-Counting measurements mainly of different architectural styles for comparison of fractal-like works of architecture with representatives of Modern architecture. Those buildings where the single measurement points in the double logarithmic graph are very close to the replacing line will be analyzed more closely with regard to a possibly underlying Fractal concept. An interesting aspect will then, however, be in how far such a concept has some influence on architectural quality and on the acceptance of the building by observers.

References

1. B.B. Mandelbrot, *Die fraktale Geometrie der Natur* (Birkhäuser, Basel [u.a.], 1991)
2. Le Corbusier, *1922, Ausblick auf eine Architektur* (Vieweg & Sohn, Braunschweig/Wiesbaden, 1993)
3. N.A. Salingaros, *A theory of Architecture* (Umbau-Verlag, Solingen, 2006)
4. H.-O. Peitgen, H. Jürgens, D. Saupe, *Chaos and Fractals – New Frontiers of Science* (Springer, New York, NY [u.a.], 1992)
5. C. Bovill, *Fractal Geometry in architecture and design* (Birkhäuser, Boston, Mass. [u.a.], 1996)
6. H. Borcherdt, *Architekten, Begegnungen 1956–1986* (Georg Müller Verlag, München, 1988)
7. M. Küper, I. van Zijl, *Gerrit Th. Rietveld, The complete works* (Centraal Museum, Utrecht, 1992)

8. C. Jencks, *The Architecture of the Jumping Universe* (Academy Editions, London, 1996)
9. R. Zerbst, *Antoni Gaudí, 1852–1926* (ein Leben in der Architektur, Taschen, Köln, 1993)
10. M. Burry, *Gaudí unseen – Die Vollendung der Sagrada Família* (Jovis Verlag, Berlin, 2007)
11. P. Portoghesi, *Nature and Architecture* (Skira editore, Milan, 2000)
12. G. Lynn, *Animate Form* (Princeton Architectural Press, New York, NY, 1999)
13. H. Maertens, *Der optische Maßstab* (Verlag von Ernst Wasmuth, Berlin, 1884)
14. K. Foroutan-pour, P. Dutilleul, D.L. Smith, in *Applied Mathematics and Computation*, Advances in the implementation of the box-counting method of Fractal Dimension estimation, vol. 105, Issue 2–3 (Elsevier Science Inc., New York, NY, 1999), pp. 195–210
15. D. Hoffmann, *Frank Lloyd Wright's Robie House* (Dover Publ., New York, NY, 1984)
16. W.E. Lorenz, Master Thesis, *Fractals and Fractal Architecture* (Vienna University of Technology, 2003)
17. W.E. Lorenz, in *First International Conference on Fractal Foundations for the 21st century architecture and environmental design*, Fractal Geometry as an approach to quality in architecture (ffractarq, Madrid, 2004), CD-ROM
18. M.J. Ostwald, J. Vaughan, C. Tucker, in *Nexus VII: Architecture and Mathematics*, Characteristic Visual Complexity: Fractal Dimensions in the Architecture of Frank Lloyd Wright and Le Corbusier (Kim Williams Books, Turin, 2008) pp. 217–231
19. M.J. Ostwald, C. Tucker, in *Techniques and Technologies Transfer and Transformation: IVth International Conference of the Association of Architecture Schools of Australasia 2007*, Measuring architecture: Questioning the application of non-linear mathematics in the analysis of historic buildings (University of Technology and AASA, Sydney, 2007), pp. 183–189
20. H.-O. Peitgen, D. Saupe, *The science of Fractal images* (Springer, New York, NY, 1988)

Picture credits

All pictures by Wolfgang E. Lorenz.

Part III
Information and Dynamics

Chapter 10
Biomimetics in Intelligent Sensor and Actuator Automation Systems

Dietmar Bruckner, Dietmar Dietrich, Gerhard Zucker, and Brit Müller

Abstract Intelligent machines are really an old mankind's dream. With increasing technological development, the requirements for intelligent devices also increased. However, up to know, artificial intelligence (AI) lacks solutions to the demands of truly intelligent machines that have no problems to integrate themselves into daily human environments. Current hardware with a processing power of billions of operations per second (but without any model of human-like intelligence) could not substantially contribute to the intelligence of machines when compared with that of the early AI times. There are great results, of course. Machines are able to find the shortest path between far apart cities on the map; algorithms let you find information described only by few key words. But no machine is able to get us a cup of coffee from the kitchen yet.

Biomimetics, being the application of biological systems found in nature to the study and the design of engineering systems and modern technology, is the promising method. However, it has to be implemented reasonably, which is argued and detailed below. Doing so, a new research field – the alliance between engineering and psychoanalysis – emerged, which is presented in this contribution. The ultimate goal of this research is to create a human-like intelligence for the control of automation systems that consist of sensors and actuators interconnected via field bus systems.

D. Bruckner (✉)
Institute of Computer Technology, Vienna University of Technology, Gußhausstraße 27–29, 1040 Vienna, Austria
e-mail: bruckner@ict.tuwien.ac.at

10.1 Research Field

The core question is: What is necessary to give machines an understanding of the real world? The postulated answer is: The integration of research findings from psychoanalysis, neurology, and engineering was the basis for the foundation of a new research area, which focuses on completely new ideas. Instead of looking for solutions for partial problems, the focus should be put on a holistic model of the human psyche describing all of its functions – up to consciousness. Even if the implementation of machine consciousness is still to come, the model development of necessary functions and requirements gives important hints for further research.

One important research task today is the evaluation of information and the extraction of relevant information. This is necessary for recognizing situations and for decision making. The complex evaluation system of humans serves as a biological archetype: through comparison with a huge number of already evaluated patterns, we can filter and classify lots of information very quickly. Psychoanalytic theory models this evaluation process with the help of emotions, affects, and drives. The human psyche needs to satisfy various requirements from different sources like the own body or the environment while it constantly facilitates equilibrium between those partly contradicting demands. A technological implementation needs to facilitate exactly these mechanisms, in order to have the same capabilities. This task poses many unintended obstacles (e.g., the definition of a technological "body"), but it is seen as the lead to future success.

Researchers face the challenge to translate domain-specific knowledge from foreign disciplines into their own. This knowledge is not edited to be understood by engineers. It has to be "translated." Very often mechanisms are described in terms of their observable function, not in terms of their functional model – as required in engineering.

10.2 Automation

From the early days on, the focus of scientific investigations in the field of automation was on communication technology. Industry quickly identified the enormous market potential of distributing intelligent control units to various fields such as energy production, energy supply, rail transportation, or aviation and space technology. This and the following section intend to review history to better understand the present and how the current research directions evolved.

The term automation was introduced in 1936 by the Ford manager D.S. Harder [1]. Since that time, its meaning has changed remarkably. In the beginning, automation only meant optimization of processes in the production of goods. In engineering today, the meaning of automation has changed to the more general notion of processes controlled by machines. Hence, in the past its major goal was optimization of mass production. Today, no matter whether human resources are additionally required or not, we talk about automation when processes are controlled by machines, though mass production is still an important goal. However, automation today is about more, about process optimization in various respects, e.g., higher quality, safer environments, more security, or better hygiene.

In previous times, the term "hand-made" has been a quality feature. Today, when talking about material goods, it refers no longer to high quality, but rather to the opposite (of course this statement does not hold for artistic craftwork). This is because machines are able to continuously deliver high quality. The quality can be only kept on high levels and standards with consolidated deployment of automation technologies and methodologies. Otherwise, micro and nanotechnology, chip technology, and many other areas of production procedures could not be implemented. Impressive examples of advantages through automation can be found, e.g., in aviation technology. Only field bus technology and the introduction of the fly-by-wire principle made modern aircrafts possible. Another example is cars. Modern upper class cars have some hundreds of embedded systems communicating with each other, drive-by-wire. Both "by-wire" principles imply that the device is no more directly controlled (e.g., with a mechanical bar), but with the use of a field bus system. The term field bus system refers to a communication and computation system where various sensors, actuators, and control units are interconnected with a communication medium, the field bus. The principle is the same as in bus systems in a common computer, but the application is outside of it, e. g. in a factory or a car, that is "in the field."

Mining companies in Europe were only able to reach the required safety level for road tunnels through interconnection with field busses. They steadily collect data and send it to control stations and central offices. In this case, data collection was not a technical problem at all. It could have been accomplished with decentralized control units or even with widespread internet connections for all sensors and actuators. However, the key advantages of field busses are twofold: first of all low prices, and second and most important their specification of profiles for application areas. Especially the second point, the definition of profiles, is a question of standardization, for which industry, industrial associations, and also governments spent lots of money in the form of grants. In this way, a broad basis emerged, on which most notably the European industry created enormous margins. Without profiles, field busses are worthless; they would be too expensive in development and still more so in maintenance. If countries that do not have strong industries in this area want to use this fundament, they need people who acquire the knowledge about profiles for them. The conception of a field bus being a small computer with long connections to sensors and actuators as well as connections to other computers is no longer enough, since field bus components will be superseded in the next years with developments in the areas of internet technology and brown ware (consumer electronic). According to Moore's law, the performance of processor chips will increase year by year. Therefore, knowledge in the field bus area does not lie in components; it lies in the functional profiles and interoperability. And still there is more to it. The second large area of knowledge capital, which makes out a field bus expert, can be seen in application-specific knowledge. The collection of billions of data points via a large field bus system is not a trivial task at all; especially if various channels are connecting sinks and sources. But the real challenge lies in finding the information behind the data – which can only be accomplished automatically due to the sheer amount of data [2].

10.3 Intelligence and Communication

Two major issues must be discussed: First, why is it now that automation experiences its phenomenal boost as an essential pillar of economy? Second, a vast number of processes can be found in nature. Which automation principles can be found there and what can biomimetical approaches offer for even smarter control units?

Purely mechanical control systems are always physical compromises. For the control of a process A to use some mechanical system B that is subjected to physical laws and constraints compromises are necessary. The differential transmission of the two front wheels of a vehicle is an example. It must be constructed to support minimal curve radius with various velocities and torques. A differential transmission of a tractor therefore looks different from one of a race car.

If we separate the flows of energy and information in a mechanical process and develop separate electronic control components (which are not subjected to physical constraints), those compromises become almost always unnecessary. Let us assume that the mechanical differential transmission together with the steering column will be replaced by propulsion engines directly integrated into the wheels (which in fact is a goal of the automotive industry, remember fly-by-wire: drive-by-wire). In this case, any desired differential transmission algorithm can be implemented with the drive-by-wire system (field bus). Such a system could also additionally be able to consider road conditions. Hence, compromises regarding control are no longer necessary. Therefore, economic benefits arise. Mechanical degradation is limited to the remaining parts for suspending the wheel. Maintenance will be eased and therefore available for lower prices. The period of warranty can be enhanced, because the failure rate can be calculated more precisely and easily. The overall energy consumption can be optimized; also the wear of tires can be reduced.

All these considerations reveal that with the help of automation and the introduction of field bus systems, production lines can be built more efficiently, in short times, for lower prices. Large portions of energy can be saved. All this was not possible in earlier times without field bus technology. It became a crucial business factor. Smart control units implemented as embedded systems together with communication technology – as learned from nature – are the motors of this trend.

Still, the explanation is missing how evolution tackled those problems. Very old creatures such as the amoeba can be treated like the mechanical systems in the above examples with respect to their control (= information processing), since they act based on physical–chemical processes, which does not allow for any separation of information and energy flow. However, evolution introduced this

Fig. 10.1 Possible abstraction layers of a computer

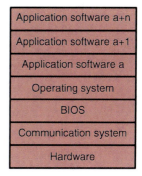

principle, e.g., already in worms (which are one example for rather old creatures that have nerves for communication). However, what about humans? We have special information communication channels (the nerves) evolved, which have been the template for the technology of field bus systems (fly-by-wire principle). The information transportation mechanisms are however somewhat differently implemented. Nerves do not exactly work like bus systems. However, this does not contradict the basic principle of separating energy and information flow.

In computers, we differentiate various abstract layers (see Fig. 10.1) from the lowest layer, the hardware (which can itself be divided into sublayers), via device drivers, operation system, up to the application software (which, again, is modeled itself in many further layers and sublayers). The same can be done with the information processing system in living creatures [4]. Nerves and neurons themselves form in this sense the hardware. Higher layers in the brain have been found by the neurologists Luria [3]. In nature, peripherals and centralized oriented information systems can be observed, pursuing decentralized control principles. Because of the actual slightly different functionality of nerves and technical communication systems, it can be anticipated that the principle of the human nervous system needs to be adopted and only implemented in principle, if complying smart automation systems should be constructed. But it can be also useful to look at control mechanisms of various animals to achieve this goal. In this way, bionics gains even more importance, especially with respect to the higher layers of information processing, which are the key enabling factors for "smartness."

10.4 Open Problems: Challenges in Research

The section should highlight that different applications for particular field bus systems require different profiles. A profile is seen as a higher level of abstraction in communication systems. On the one hand, all profiles have distinct properties and justification for particular applications. On the other hand, because profiles are more or less independent of the lower layers (in particular the lowest one, the hardware), the question arises whether it is possible to harmonize the hardware of different field bus systems. Today's discussion is: Can hardware be found that is economically applicable in many different bus systems, allowing for different implementations of the higher layers? One candidate could be Ethernet, which is under constant development and, through its widespread deployment in the PC world, economically priced. Many indicators point at a possible application of Ethernet for field busses, but in the history of field bus technology many technically reasonable solutions have been presented, which were then no longer followed due to economical reasons. The discussion is still open, although the trend goes to Ethernet (and its hardware) to be utilized on a broad basis for the lower layers in field bus systems.

The last point also refers to wireless networks. The developments in this area are also still in progress, especially focusing on energy efficiency while preserving communication abilities. The duration of such nodes reached already some years, but shall be enhanced up to five or more years [5] to allow for cost-efficient applications, as in transportation [6].

Research and development in field bus technology aside the harmonization aspect concentrates mainly on four topics: reduction of installation and maintenance costs, safety, security, and interpretation of the enormous amounts of data. The first three topics are mainly unrelated to biomimetics, while the latter is a clear case of a bionic application and will be dealt with in the next section in more detail.

The problem of reducing costs, especially for installation and maintenance, is of great significance in building automation, since conditions are most dramatic. The physical placement of nodes is usually performed by very cheap, semiskilled workers, and the number of nodes is in a dimension of several thousand up to many ten-thousands of nodes. Another problem of integrating and maintaining such a number of components in an economic way on a PC is left to the integrator. However, this is not only a realization question to an engineer, but also a scientific challenge. This is because modern technology did not find a suitable solution which satisfactory supports the integrator with this problem on the construction site.

Safety (functional safety targets ensuring normal operation even if subsystems fail) is a topic that was rarely tackled in automation and sometimes even in an amateurish way. An example should demonstrate those first, inadequate considerations. It was the time when safety relevant electronic circuits should be realized with micro processors: it was demanded to build up the circuits redundant in a way that one circuit is allowed to contain a micro processor, but the second needs to have a discrete-built circuit. Times changed this fundamentally, but still safety is often avoided since it is seen to imply higher costs, sometimes more than double costs. However, experts know this is absurd, since safety is not about doubling hardware or software, but functionality, which results in additional costs of the product often in a range of 5–10%.

The case with security is even worse (protecting a system against attackers). In automation, this topic is widely untouched; some proponents even say that a firewall at system border is sufficient. It is certainly not, e.g. in many critical infrastructures, attacks come from active or fired employees, thus from inside. For mechanical systems, this was hardly a problem, but in case of electronic systems it is. Imagine connecting a small hidden device to an internal field bus, which is remotely controlled to terrorize a company. Although the topic is of high importance, in many areas such as industrial automation or automation in transportation this topic is kept behind the scenes these times. Only in building automation some considerations have been undertaken since an attacker could annoy the working or living people in the building a lot. But, as mentioned earlier, it is of no concern today. Research results are present in small number (see [16]) but it is questionable if they will find integration in practice or if they will be further investigated. Maybe other solutions will be found, which are more economical.

After this, the last and maybe most interesting topic is reached: How can the enormous amounts of data in a building, e.g. an airport or a shopping center, be usefully interpreted without the need for massive support by humans? This question already implies that a system is sought that can substitute the capabilities of humans in this respect. But what respect is it when referring to building automation? The applications are HVAC systems (heating, ventilation, air condition), illumination, security systems, and – in the vision of many researchers – context aware systems for many applications to increase safety, security, comfort, energy efficiency, etc. The requirements on a meta level for all applications are similar: perceiving information and interpreting this information with respect to the context of the system. But what is the context? The context is a theoretical construct by humans. Therefore, for a machine to be able to derive the context, it is necessary to give it the same perceptual function as a human. This problem can also be only tackled with using bionic's principle to study nature's solutions. The next section will describe the decision processes that finally lead to the development of the bionic research area at the Institute of Computer Technology of the Vienna University of Technology.

10.5 Intelligence of Bionic Systems

Before presenting the current state of the model development, it is necessary to explain the motivations why this development was essential.

10.5.1 Hierarchical Model Conception

The various approaches of artificial intelligence (AI) over time have been summarized in [3]. Four generations have been identified: symbolic AI, statistical AI, behavior-based AI, and recently, emotional AI. The article very well describes how researchers started to build and understand functions of the human mental apparatus starting with neurons in a bottom-up design methodology, or even with data-driven statistical analysis. Very much emphasis has been put for a considerable time period – sometimes including today – on a behavior-based design methodology, meaning that devices like robots have been built that behave similar to (e.g., move

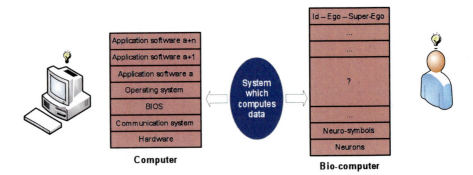

Fig. 10.2 Possible abstraction layers of a computer (to the *left* the artificial device, to the *right* a biological model conception)

like) humans or animals. The term intelligence, however, has never been precisely defined – it has not even been used, except for the title "AI."

However, if engineers are to design machines that should behave "intelligently" as animals or humans, it has to be dealt with the definition thereof. So, what is intelligence? The question can only be seriously answered, if we are able to understand the mental apparatus. It is not enough to know about the behavior, but it is mandatory to have a unitary and comprehensive functional model of it.

For the below presented model development, a number of boundaries has to be defined. On the one hand, appropriate modeling principles and methods from computer engineering, communications engineering, and automation are to be used. Abstract layered models are used there, which are developed in a top-down fashion.

On the other hand, for comparing the platforms, we utilize the definition of a computer after which it is a data manipulating, storing, and transferring device. In this sense, an artificial computer and a biocomputer can be compared (Fig. 10.2). There are many abstract layers defined in an artificial computer in the left part (from hardware to application software) – which can be also imagined in the biological model that is shown in the right part, where many layers are not yet defined due to lack of knowledge. Neurologists also second this structure [3].

Figure 10.2 expresses moreover also the monist conviction – also along with the natural science view of the world – that the brain is a control system based solely on the laws of physics, with no principally unexplainable mechanisms, no mystics [18, 19 p.51].

10.5.2 Statistical Methods

A large number of feedback loops of all kind both physiologically and mentally (hence, in all layers of Fig. 10.2) are active in humans (Fig. 10.3). It can be easily imagined that specific propositions about particular feedback loops in such

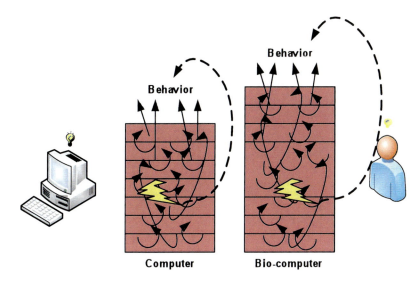

Fig. 10.3 Effect of layers: different behavior of the whole process

an arrangement can only be stated, if there exists a concrete conception about the whole process. In the layered model, two completely different errors in different layers can result in the same behavior. Or, a substantial error in one layer can be compensated by actions in other layers.

So, if the behavior of the whole process is under investigation from the outside, observed phenomena can only give indications about feedback loops inside without having an accurate description of all feedback loops.

For a better depiction another small example is given (Fig. 10.3, left). If there is an instability or error in the operation system, the observer who works with the application software may see only that crashing and believes the error occurred there. This is due to lack of knowledge about the correlations.

This description should elaborate why computer engineers (and chip designers) do not use statistical methods of behavior for synthesis. They need an explicit model description. However, AI often uses behavior-based methods, based on statistical analysis for synthesis as can be seen in the example of [7] – a principle, which has to be questioned.

10.5.3 Definition of Intelligence

The terms intelligent and intelligence have to be used with care. Some two or three decades ago, there have been severe discussions about which processor is the more intelligent one (they spoke about architecture and computing power). However, it was understood soon that this is the wrong kind of question. Computing power

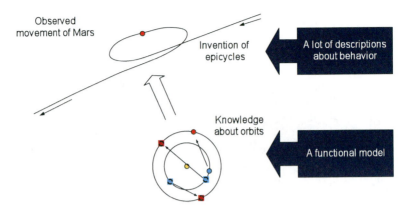

Fig. 10.4 The earth as the center of the universe

cannot be defined in general but only application specific in terms of problem solving. Thus, if a hierarchical model as in Fig. 10.2 is assumed, such kind of question can only be applied to a single layer. Or the range of applications has to be entirely defined, which are performed by the computer. So, to define intelligence one has to be aware to which layer he is referring to. For the synthesis of a model, a computer designer needs a conception for every single abstract layer.

10.5.4 Choice of the Right Model

Here is an example that shall point out the importance of the right point of view. In medieval times, it was assumed that the earth is the center of the universe. The description of Mars' orbit (Fig. 10.4) was therefore mathematically problematic and only possible in approximation with the help of epicycles (ptolimeic world view).

The breakthrough came in 1543 as Copernicus saw the sun as the center – so all orbits could be easily modeled as ellipses.

Transferred to observed human behavior, this means that descriptions based on an erroneous (or from the wrong point of view) model tend to give misleading conclusions. For synthesis of a bionically-inspired model, one needs the right model.

10.5.5 Top-Down Methodology

For computer design or chip design, it has to be distinguished between behavior and function. This means that if a particular potential behavior is sought, one needs a model of the device under development, which can then be iteratively enhanced to reach the desired behavior.

Furthermore, for model development, it is mandatory to use the top-down methodology, because otherwise the optimal circuit design cannot be found. If, for example as in [7], two inconsistent behavioral models would be used as basis, it seems obvious that inconsistencies arise. Here the authors understand top-down as being the design process that starts on the top most abstract functional level – the main function, which is then divided into subfunctions and modules in the consecutive next lower hierarchical abstract layers until a specific description is reached, which can be actually implemented, and on the lowest layer synthesized in real hardware.

10.5.6 A Unitary Model

There exists an abundant amount of psychological theories as has been tried to review in [7]. In this book, the attempt has been started to collect all psychological studies. However, they are taken and publications are cited without having checked their compatibility (interoperability). At least, it is not reported. However, interoperability is essential in the area of interdisciplinary science when taking other theories as template.

In case of psychological schools, they are often based on different premises. The approach to theory-making of psychology cannot be compared with natural science practice where it can be seen as major goal to eliminate inconsistencies. Additionally, in the psychological area, it is very hard or even impossible to formally proof a theory. Therefore, more empirical procedures have been applied. However, on the one hand, if the goal is to compile the mental apparatus in the bionic sense with methods from computer engineering, one cannot simply accept inconsistencies, but needs to search for solutions or particular application cases. If, on the other hand, results or statements from different psychological schools are used, which are not interoperable, the resulting model will be only a patchwork.

10.5.7 Differentiation Between Function, Behavior, and Projection

In previous sections, we already distinguished between behavior and function. However, there is one additional distinction necessary: projection. When looking at quite famous developments such as the CB2 (a robot looking like a human child of 4 years) [W1], we realize the way how reports about it are formulated. A lot of analogies to children are endeavored. However, from a functional point of view, a robot has nothing in common with a human because it simply lacks the functionality of the psyche. Behavior created from the robot is driven by (very sophisticated, of course!) computer algorithms that work very different than the human psyche. The one thing that makes people call it human-like is its appearance. Thus, observers project human behavior into the robot. It is the same as with a doll. Even if a number of actuators remarkably affect CB2's facial expressions, there is no machine emotion behind, only observers perceive the expression of an emotion. Hence, it does not smile because it is amused; there exists no concept of amusement in the machine or doll. Once again, it simply lacks the psyche to generate any kind of human-like behavior. It is only the observers that project human-like emotions and behavior into the robot.

10.5.8 Indispensible Interdisciplinarity

In science, there exists the principle that all relevant scientific results worldwide have to be incorporated to serve as the state-of-the-art. However, when developing human-like systems, it seems that this principle has been avoided since psychoanalysis – which is as their core competence concerned with the topic for more than 100 years – has not been considered.

To emphasize this point and weaken arguments toward possible self-studies, another hint is necessary: the education of an engineer at a university in Europe lasts for about 6 years, a psychoanalysts needs about 9 years [W2], [W3]. This means that an engineer, on the one hand, will probably not have time to undergo this education, and on the other hand, conducting self-observations without such education will probably not lead to any useful template for implementation.

Having a look at international working groups on the issue of Artificial General Intelligence, it can be noted that hardly either psychoanalysts or neuropsychoanalysts are actually involved. Only some suggestions to bring together psychoanalysis and engineering have been made. The first investigations in this direction were presented in [4] and [8–12].

In other parts of AI, it has become common practice that engineers interpret and utilize psychological literature. But then an engineer would be quite astonished if a psychologist or even psychoanalyst came to the idea to interpret technical research articles. The conclusion is clearly that in such kind of working groups interdisciplinary work has to be more emphasized [13–15].

10.6 The Psychoanalytical Model

Finally, this section presents the developed model. This model is intended to be used in decision units of various technical devices like robots or automation units. It functions according to the psychoanalytical description of the functionality of the human psyche. Therefore, it incorporates terms such as drive, wish, emotion, affect. The following description is very dense and should give the reader only an overview of the topic.

10 Biomimetics in Intelligent Sensor and Actuator Automation Systems 215

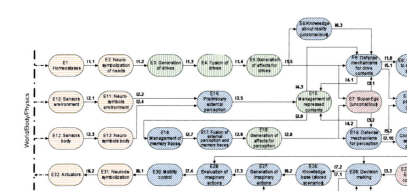

Fig. 10.5 Functional model of the psychic apparatus

A functional model (Fig. 10.5) has been developed, which describes with the help of psychoanalytical concepts how a motivational moment generates a wish, then how decisions are taken, actions are planned, and finally conducted (a detailed description can be found in [17]).

It is the task and function of the human mental apparatus to synthesize the demands of three instances. These instances (Fig. 10.5) are the drive demand (E3), which bases on bodily-physiological requirements (E1), the reality demand, composed of internalized knowledge about the facts of the outer world, its possibilities and bounds (E9), and the subjective consequences developed from perception of the outer world (E14). The third instance represents the demands of the Super-Ego (E7, E22), which is composed of socio-cultural founded bids and bans, i.e. rules.

These three instances and the consequences of their demands for the mental apparatus are to be comprehended. A physiological, hormonal imbalance that develops in an organ (E1) triggers via neuro-symbolization of that organic requirement (E2) a drive tension (E3), whereby the drive tension is the first psychic representation of bodily circumstances. Life-sustaining drives are mixed with aggressive drive tendencies (E4). Their content in the form of thing presentations is further transported (I1.4) and gets rated by an affect of more or less unpleasure (E5). Affect together with (thing) presentation form the drive, whereby the affect is the assessment thereof.

Affect and thing presentation together are carried (I1.5) to the defense mechanisms (E6), which with the rules about bids and bans from the Super-Ego (E7), resulting fear, and under the internalized reality demands (E9) decide, if and in what form affect and presentation are handed over to potential conscious processing in the Ego.

Affects and presentations, which are entirely defended through the defense mechanisms, as they are not allowed to get conscious or even preconscious, remain in the Id part of the mental apparatus in a container for repressed mental contents (E15).

Affects and presentations that gained access (in what any kind of form) to the Ego become processed in terms of the secondary process (E8). This means that thing presentations become additionally conjunct with word presentations. In that form thing presentations can be ordered and assessed logically, meaning they become consistent with temporal and spatial conceptions.

As has been already stated, the work of the mental apparatus is influenced by drive demands (E3) arising from bodily needs, perceptions of objects, and circumstances of the outer world (E14). The outer world is represented in sensor data of two kinds: one targeted to the environment (E10), and one solely targeted to the own body (E12). These raw sensor data are transformed into neuro-symbols representing body and environment (E11, E13). These are the content of unconscious perception before any kind of psychic processing or assessment. The perceived (thing) presentations come into contact with repressed mental content and together activate memories (E16). As a result, a combination of perception from the outer world and memory is constructed (E17). This means that not some kind of copy of the outer world is psychically processed, but a subjectively and individually assessed and associated presentation thereof. Just like the drive tensions, the presentations of the outer world produce affects (E18). Presentations and affects on this level originating from outer perception and demands are similarly treated by defense mechanisms (E19) as it has been described for drive tensions. Again, perception content that passed the defense mechanisms and therefore gained access to the Ego (in what any kind of form), become processed in terms of the secondary process (E21).

At this level for the first time, preconscious or conscious inner perception (E20) of drive tensions and perceived content from the outer world is possible. Inner perception includes also preconscious/conscious perception of feelings and affects that are related to the combined thing/word presentations.

The thing/word presentations and related affects are carried (I1.7, I2.11, I5.5) to three further processing entities: the preconscious Super-Ego (E22), the decision unit (E26), and the attentive outer perception (E23). The preconscious Super-Ego in principal contains the same content as the unconscious one. However, its content can be accessed consciously and has influence on the decision unit (E26). It controls how the wish – resulting from drive tension after secondary processing – is to be treated. In other words: if and how wish-fulfillment can be achieved.

Both, the attentive outer perception (E23) and the memorized facts about circumstances of reality (E25) affect the reality check (E24), which tells the decision unit what in terms of wish-fulfillment can realistically be achieved and what not – without moral assessment.

After drawing a preconscious/conscious decision that wish-fulfillment is to be achieved (E26), with the help of memorized scenarios (E28) potential action plans are composed (E27) and assessed (E29). The decision for conducting a particular action plan is thereby mainly influenced by feelings (I5.5) resulting from inner perception (E20).

Finally, the action plan is decomposed into instructions for motility control (E30), which are neuro-desymbolized (E31), meaning that mental content is translated into physical signals, which control the actuators (E32). The feedback loop is closed via the sensors that perceive the actual affect on the own body and also the actual affect on the environment of the conducted actions.

10.7 Conclusion

Communication and automation technology have evolved from simple mechanical systems over systems with simple automation and communication abilities to technologically complex solutions that implement all levels of the ISO/OSI communication model and even more abstract levels above and perceive and affect their environment in particularly very advanced ways. The academic questions to be solved in this context have changed from classical electrical engineering issues to problems, which are more related to computer science since the emphasize on the informational part in automation systems continually increases. The demands and requirements for automation systems are steadily changing. In the context of smart future automation networks, it is of major importance to add smart functionality to devices. In the long run, only human-like capabilities will allow for automation of many fast, dangerous, and demanding processes. This goal can in our perspective only be achieved with interdisciplinary scientific efforts between engineers and psychoanalysts in a bionic fashion.

References

1. *Brockhaus encyclopedia in 24 volumes (German)*, Vol. 2, 19th edn., p. 409, 1987
2. D. Bruckner, Dissertation thesis, Vienna University of Technology, 2007
3. A.R. Luria, *Basic Books*, The Working Brain: An Introduction To Neuropsychology (1992) ISBN 978–0465092086, Basic Books, New York
4. D. Dietrich, G. Fodor, G. Zucker, D. Bruckner, *Simulating the Mind – A Technical Neuropsychoanalytical Approach* (Springer, Wien, 2008) ISBN 978-3211094501
5. S. Mahlknecht, S. Madani, On Architecture of Low Power Wireless Sensor Networks for Container Tracking and Monitoring Applications, IEEE Int. Conf. Ind. Inform., 353–358 (2007)
6. S. Madani, S. Mahlknecht, J. Glaser, Clamp: Cross Layer Management plane for low power Wireless Sensor Networks, in *Frontiers of Information Technology*, Islamabad, Pakistan, 2007, p. 9.
7. Sh. Turkle, *Artificial Intelligence and Psychoanalysis: A New Alliance*, ed. by S. R. Graubard (MIT Press, Cambridge, MA, 1988)
8. D. Dietrich, Evolution potentials for fieldbus systems, in *IEEE International Workshop on Factory Communication Systems WFCS 2000*, Instituto Superior de Engenharia do Porto, Portugal, 2000
9. G. Russ, Dissertation thesis, Vienna University of Technology, 2003
10. C. Tamarit, Dissertation Thesis, Vienna University of Technology, 2003.

11. B. Palensky, (née Lorenz), Dissertation thesis, Vienna University of Technology, 2008
12. C. Rösener, Dissertation thesis, Vienna University of Technology, 2007
13. P. Palensky, D. Bruckner, A. Tmej, T. Deutsch, Paradox in AI - AI 2.0: The way to machine consciousness, in *CD Proceedings of IT Revolutions*, Venice, 2008
14. M. Ulieru, R. Doursat, Emergent Engineering: A Radical Paradigm Shift, Int. J. Autonomous Adap. Commun. Sys. (2009)
15. A. Kirlyuk, M. Ulieru, IT Complexity Revolution: Intelligent Tools for the Globalised World Development, in *CD Proceedings of IT Revolutions*, Venice, Italy, 2008
16. W. Granzer, W. Kastner, G. Neugschwandtner, F. Praus, Security in networked building automation systems, in *Proceedings of the WFCS*, 2006, pp. 283–292
17. R. Lang, Dissertation Thesis, Vienna University of Technology, 2010
18. R. Dawkins, *The God Delusion* (Random House UK, 2007) ISBN 978–0552773317
19. M. Solms, O. Turnbull, *The Brain and the Inner World* (Other Press, LLC, 2002) ISBN 978–1590510353

Web-Links, (All accessed March 2010)

1. http://www.youtube.com/watch?v=bCK64zsZNNs
2. WAP: Wiener Arbeitskreis für Psychoanalyse, http://www.psychoanalyse.org
3. WPV: Wiener Psychoanalytische Vereinigung, http://www.wpv.at

Chapter 11
Technical Rebuilding of Movement Function Using Functional Electrical Stimulation

Margit Gföhler

Abstract To rebuild lost movement functions, neuroprostheses based on functional electrical stimulation (FES) artificially activate skeletal muscles in corresponding sequences, using both residual body functions and artificial signals for control. Besides the functional gain, FES training also brings physiological and psychological benefits for spinal cord-injured subjects. In this chapter, current stimulation technology and the main components of FES-based neuroprostheses including enhanced control systems are presented. Technology and application of FES cycling and rowing, both approaches that enable spinal cord-injured subjects to participate in mainstream activities and improve their health and fitness by exercising like able-bodied subjects, are discussed in detail, and an overview of neuroprostheses that aim at restoring movement functions for daily life as walking or grasping is given.

11.1 Introduction

Injuries or diseases can interrupt the conduction of action potentials in the neural system. Depending on the kind and severeness, this may lead to complete or partial loss of control of the muscles of the lower and/or upper extremities.

The application of electrical stimulation in a rehabilitative setting was initiated in 1961, when W.T. Liberson, a physical rehabilitation specialist and medical researcher, developed a heel switch-triggered personal electronic stimulator device to correct foot drop [1]. Functional electrical stimulation (FES) aims to generate movements or functions which mimic normal voluntary movements and so restore

M. Gföhler (✉)
Research Group for Machine Elements and Rehabilitation Engineering, Institute of Engineering Design and Logistics Engineering, Vienna University of Technology, Karlsplatz 13, 1040 Vienna, Austria
e-mail: margit.gfoehler@tuwien.ac.at

the functions which these movements serve. Devices that are delivering FES are a type of neuroprosthesis.

Reactivation of skeletal muscles and hence movement functions by FES may have impact on general life, reduce secondary health problems, and increase overall quality of life. Specific training with FES can cause significant improvements of the cardiovascular and pulmonary systems, reduce atrophy of skeletal muscle, increase bone density, and also lead to mental benefits ([2, 3]; Faghri et al. 1992).

11.2 Principle

Naturally, the signal for muscle contraction is generated in the central nervous system (CNS). This signal is propagated to and along the peripheral nerve and via the synapsis transferred to the muscle, where it induces the contraction. If this natural muscle activation process is interrupted by a lesion, the activation signals from the CNS cannot reach the muscles and consequently the muscles are paralyzed. FES is a method to artificially generate an activation potential in the peripheral nerve. A stimulator sends a stimulation pulse to the electrodes, and an activation potential is generated in the peripheral nerve and propagated to the muscle in the same way as in the physiologically intact body.

Figure 11.1 shows the main components required for a neuroprosthesis based on FES. Central element of an FES-based neuroprosthesis is the FES controller, which receives command signals and sensory input from artificial and natural sensors and controls the electronic stimulator. The stimulator generates stimulation pulses and induces muscle actuation via electrodes.

11.3 Actuation

The goal of FES is to stimulate the paralyzed muscles in as natural manner as possible. This requires that the muscles are activated selectively and produce reproducible graded forces. However, many poorly controllable factors related to neuromuscular anatomy and electrode placement make these goals difficult to achieve.

Muscle activation by means of electrical stimulation usually aims at generating an activation potential in the peripheral motor nerves that innervate the muscle, presuming that the peripheral motor nerves are not damaged. Also reflexes can be elicited by electrically stimulating the afferent nerves. Muscle fibers themselves are in principle electrically excitable but require very high stimulation intensities.

The stimulation signal for FES is generated by a programmable stimulator and transferred to the electrodes which transduce electron current into ionic current in the tissue. If the depolarization is strong enough, an action potential is induced in the nerve and propagated along the nerve fiber. This activation potential is then chemically transferred to the muscle fibers via the synapsis and induces muscle contraction and consequently the tendon force. The activating function $f(x,t)$ is

11 Technical Rebuilding of Movement Function

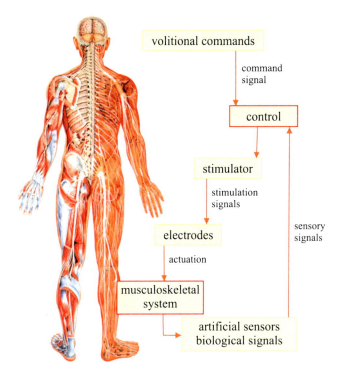

Fig. 11.1 Main components of an FES-based neuroprosthesis (figure of the human ©2010 3B Scientific GmbH, Germany)

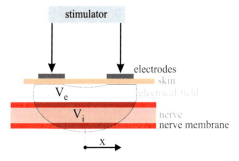

Fig. 11.2 Schematic of electrical field distribution in the tissue under surface electrodes

defined at each instant of time t as the second derivative of the extracellular potential V_e in direction x along the nerve fiber (Fig. 11.2) [4]:

$$f(x,t) = \frac{\partial^2 V_e(x,t)}{\partial x^2}.$$

11.3.1 Stimulation Signal

The stimulation signal usually consists of a train of biphasic rectangular current pulses with a frequency of between 0 and 100 Hz. Too low frequencies below the critical fusion frequency may lead to rippled muscle force output, and too high frequencies may increase fatigue [5]. For some applications, voltage-controlled pulses are used instead of current-controlled pulses; these are more easy to control but the current is the critical parameter that has to be above threshold level to depolarize the tissue and generate an action potential. The advantage of current-controlled stimulation is that if the resistance of the skin increases due to electrode drying out or sweating, the constant current stimulator will adjust automatically, whereas in a voltage-controlled stimulator the current flowing through the electrode has to be measured and the voltage adapted accordingly to achieve constant stimulation conditions. On the other hand, if an electrode loosens from the skin, the current density flowing through the remaining small contact area may increase to the level where skin damage can occur in a constant current stimulator. Generally, charge balanced pulse types are used so that no net charge is introduced to the body.

Parameters of the stimulation signal that influence the muscle force output are stimulation intensity and pulse frequency. The stimulation intensity can be varied by pulse amplitude and pulse duration, which is limited by the signal's frequency. Figure 11.3a shows a measured isometric recruitment curve (IRC), the relation between stimulation intensity (the pulse duration is varied at constant pulse amplitude) and isometric muscle force. If the stimulation intensity is higher than a threshold value, the force increases almost linearly until saturation is reached. Figure 11.3b points out that the isometric muscle force increases with stimulation

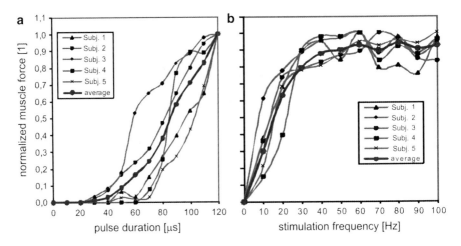

Fig. 11.3 (a) Isometric recruitment curve IRC, (b) relation between stimulation frequency and isometric muscle force, results of measurements on the Quadriceps muscles of five male paraplegic subjects and average [6]

frequency. The maximum force is reached at about 30 and 100 Hz for slow and fast contracting motor units, respectively. For frequencies above 50 Hz, muscle fatigue, which is a severe problem in FES applications, increases rapidly [5].

Muscle composition is changed in paralyzed muscle due to inactivity. The percentage of fast contracting, fast fatiguing muscle fibers increases, which is one reason for quick fatigue in paralyzed muscle. These inactivity-associated muscle changes can at least partially be reversed by FES training (Mohr et al., 1997).

In the case of physiological activation, first the thin, slow contracting motor units of the muscle are activated, and when higher forces are needed, bigger, fast contracting motor units are subsequently added. Similarly, the stimulation frequency is low at the beginning and raised for higher forces. The activated motor units are distributed over the muscle. In the case of artificial stimulation, the recruitment order is reversed to so-called inverse recruitment [7]. This means that big, fast contracting and fast fatiguing motor units are activated first. Additionally, motor units in the region of the muscle where the electrical field is stronger are activated first. Therefore, some parts of the muscle might be active while other parts are totally inactive.

11.3.2 Electrodes

Electrodes build the interface between the neural system and the technical device of the neuroprosthesis. A variety of interface concepts have been developed, ranging from simple wires to complex microsystems with integrated electronics. In general, selectivity increases with invasiveness.

11.3.2.1 Surface Electrodes

Surface electrodes are attached to the skin above a nerve or motor end plate. Their advantage is that they are noninvasive and easy to use. Disadvantages are the high influence of the electrical resistance of the skin and other tissues between electrode and nerve with respect to the distribution of the electrical field and the geometrical restrictions. It is impossible to reach deep-lying muscles without also stimulating overlying superficial muscles. High stimulation intensities are necessary, and it is difficult to predict which portions of the muscle are reached by the electrical field. One way to improve selective activation is to dynamically switch the cathode between sets of small transcutaneous electrode elements. Recently, novel embroidered electrodes have been used [8].

11.3.2.2 Subcutaneous Electrodes

Intramuscular electrodes are fine wire electrodes that are either inserted directly through the skin as percutaneous electrodes or tunneled subcutaneously. Percutaneous electrodes are less invasive than fully implanted electrodes, but positioning is difficult and relative movements can occur during movement. In addition, stress points occur where the wires cross the skin and at fascial planes between muscles. Frequent bending at these points can cause the wire to break, and the wire from the electrode comes through the skin providing a path for infection. Therefore, percutaneous electrodes are rarely used for long-term systems.

Epimysial electrodes are surgically placed on the muscle near the motor point. They consist of disk-shaped metals with a polymer shielding the surface away from the muscle.

11.3.2.3 Nerve Electrodes

Nerve electrodes are placed directly at the nerve – either adjacent, encircling, or intraneural.

Extraneural cuff electrodes consist of an insulating tubular sheath that encircles the nerve and contains two or more electrode contacts at their inner surface that are connected to insulated lead wires. The electrodes distributed around the circumference of a peripheral nerve are intended to activate different populations of axons. Cuff electrodes are easy to implant but may lead to nerve damage if their size is not well adjusted to the nerve diameter.

Intraneural electrodes are placed either longitudinally (Longitudinal Intra-Fascicular Electrode, LIFE [9]) or transversally (Utah Slanted Electrode Array, USEA [10]) in the peripheral nerve endoneurium and have higher recording selectivity and signal-to-noise ratio than extraneural electrodes. The LIFE is used for neural recording or stimulation small subsets of axons within a nerve fascicle. Typical records from LIFEs show multiunit activity where it is sometimes possible to resolve single units. LIFEs with $10\,\mu m$ thickness and 50 mm in length have been realized using thin-film microfabrication techniques on polymer substrates, which also makes them more flexible and mechanically compatible [11]. The USEA is a silicon-based, three-dimensional structure consisting of a 10×10 array of tapered silicon electrodes that project out from a $4\,mm \times 4\,mm$ substrate that is transversally inserted into the peripheral nerve for neural recording or stimulation. The lengths of the electrodes are graded from 0.5 to 1.5 mm along the length of the array to ensure that when it is inserted into a peripheral nerve, the electrode tips uniformly populate the nerve.

Sieve electrodes consist of a matrix of holes that is positioned at the end of a nerve. Ideally, the axons of the nerve will grow through the holes and build electronic contacts. Sieve electrodes have not yet been tested in human applications [11].

11.4 Stimulators

External stimulators: several multichannel programmable devices with analog and digital input and output lines are commercially available.

A commercialized *implantable* device is the eight-channel receiver-stimulator IRS-8, which receives power and control via an external close-coupled radio frequency signal. It is used in the Freehand system® for active grasp and release. Based on the IRS-8, two implantable stimulator–telemeter systems (IST) have been developed which have additional input lines for sensory signals.

Loeb et al. [12] developed a fully *implantable wireless microstimulator*, the BION ("bionic neuron," 2 mm diameter ×16 mm long). Multiple BIONs can be injected through the barrel of a hypodermic needle near the nerve or neuromuscular junction of interest. Each BION receives power and digital commands from a telemetry link and delivers current pulses of the requested duration and amplitude via electrodes that are mechanically fixed on either end of its elongated capsule.

11.5 Control

11.5.1 Modeling/Simulation

For simple tasks, muscle stimulation patterns are developed by combining clinical experience with trial and error, but it is difficult to find smooth and energy efficient movements with trial and error because of the dynamic interactions between the segments. For more complex movements, muscle stimulation patterns have to be determined mathematically by establishing a dynamic model of the musculoskeletal system. This model usually consists of rigid body segments that are linked by joints and the musculotendon actuators. Due to the complexity and variety of the biological system parameter, identification is a main problem. Many parameters are difficult to access in vivo, and there are big differences between subjects. Additionally, in the case of physically disabled subjects, changes in muscle structure occur depending on the injury. Recently, MRI techniques have brought advancements in estimating musculoskeletal data of individual subjects.

To determine the impact of electrical stimulation on a movement, the resulting muscle force has to be determined. Muscle models with varying complexity are available [13]. Generally, the muscle force generation is divided into two processes, activation and contraction dynamics, as shown in Fig. 11.4. Both activation and contraction dynamics act as a low pass filter with the output responding slower and more smoothly than the input [14]. Muscle activation corresponds to the

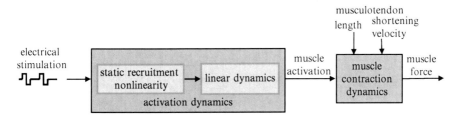

Fig. 11.4 Schematic of muscle activation and contraction dynamics

Ca-concentration and is described by the linear activation dynamics in the case of artificial activation by electrical stimulation. The static recruitment nonlinearity additionally accounts for the impact of stimulation intensity and stimulation frequency on muscle force according to the relations shown in Fig. 11.3. Muscle activation is slower and deactivation is faster in electrically stimulated muscle in comparison to physiological activation. From measurements on paralyzed leg muscles, a rise time of 108 ms was determined for 0–70% of maximal activation, and a fall time of 65 ms for 100–30% [15]. Muscle contraction dynamics describe the generation of force by activated contractile elements and basically shows the same behavior in healthy and artificially activated muscle. A muscle's force at each instant of time is a function of the instantaneous musculotendon length and shortening velocity, the tetanic muscle force, and muscle activation.

The muscle forces act on the body segments. The behavior of the musculoskeletal system is described by the equations of motion which are derived from the Newton–Euler equations. A system with n degrees of freedom (joint angles) has n equations of motion, which can be represented in vector form:

$$[A].\underline{\ddot{\theta}} + [B].\underline{\dot{\theta}} + [C].g = \underline{M},$$

where [A] is the $n \times n$ mass matrix, $\underline{\theta}$ is the vector of the system's n degrees of freedom, [B] is the gyroscopic matrix including centrifugal and coriolis terms, $\underline{M}$ is the vector of joint torques directly due to muscle forces, and the term [C].g represents the torques due to gravity. As [A] is generally a full matrix, a muscle acting on one joint can accelerate all other joints of the system, this is called dynamic coupling.

To generate a defined movement by FES, it has to be determined which of the stimulated muscles have to be active in which phase of the movement and at which level. A forward dynamic model (Fig. 11.5) is used to calculate the resulting movement trajectories from muscle stimulation. Muscle forces and musculoskeletal geometry give joint torques, then the equations of motion are used to determine the joint angular accelerations, double integration finally gives the joint angles.

If the desired kinematics is known, inverse dynamics (Fig. 11.6) can be used to determine the optimal timing of the muscle forces. The joint torques are calculated with the system's equations of motion. As each joint is usually spanned by more than one muscle, there is a distribution problem when calculating muscle forces from

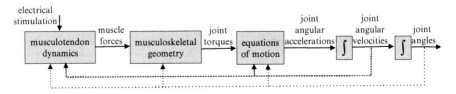

Fig. 11.5 Forward dynamic model

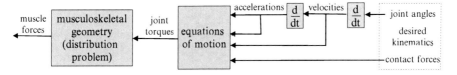

Fig. 11.6 Inverse dynamic model

joint torques. To split up the total torque to the single muscles, static optimization has to be applied. A commonly used performance criterion is to minimize muscle stress, squared, summed across all muscles [16]. As the static optimization does not consider musculotendon dynamics, the resulting muscle force trajectories may be irrealistic because a muscle cannot develop a force instantaneously.

To determine optimal stimulation patterns, muscle activation dynamics have to be considered. This is possible using a forward dynamic model where stimulation patterns are the input and the resulting movement is the output. A performance criterion applicable to the entire task has to be defined and forward dynamics combined with dynamic optimization methods to determine the optimal stimulation patterns. Parameter optimization methods as described in [17] are suitable to solve this nonlinear optimization problem, though these methods cannot differentiate between local and global maxima. Statistical methods as simulated annealing or genetic algorithms are likely to find global maxima but have the disadvantage of high computational expense. An alternative for estimating muscle forces and muscle excitations that requires three orders of magnitude less CPU time than parameter optimization is neuromuscular tracking [18].

11.5.2 Control Systems

Designing a control system that tracks a joint trajectory or torque profile by regulating the timing and levels of the electrical stimulation delivered to the muscles is a challenging problem due to the nonlinear response of the muscles to electrical stimulation and the complexity and redundancy in the musculoskeletal system.

Electrical stimulation may be delivered through either open or closed loop control systems. The FES controller attempts at taking over control tasks from the natural sensorimotor system and interfaces with the natural system at stimulation output and depending on the control approach also at command input.

In *open loop* control systems (Fig. 11.7a), the controller determines the muscle stimulation according to the desired movement trajectory, and no information about the actual trajectory is fed back to the controller. The performance of open loop systems was found unsatisfactory for the generation of accurate movements, because external disturbances as obstacles or internal disturbances as variations in muscle force generation (e.g., due to fatigue) or inaccuracies in the model of the musculoskeletal system have impact on the actual movement trajectory.

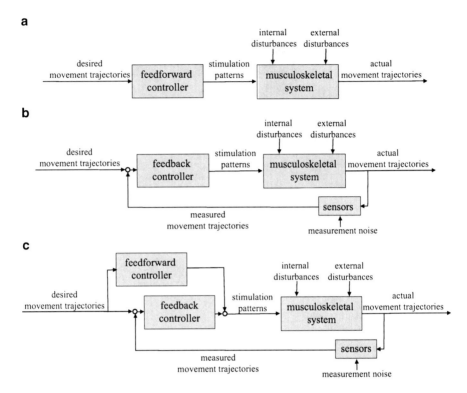

Fig. 11.7 Schematic of (**a**) open loop control, (**b**) closed loop control, and (**c**) hybrid control with both a feedforward and a feedback controller

For more complex or accurate movements, *closed loop* control (Fig. 11.7b) has to be established. During normal voluntary movements, the CNS receives sensory information on muscle-, tendon-, and cutaneous forces for neurophysiological control. In closed loop control systems, electrical stimulation is being initiated by the user's command and then modified based on some feedback measurement such as force or position. With closed loop control, the delivery of electrical stimulation is continuously modulated to control the parameter being measured by the sensors. The benefits of closed loop control are obvious, but closed loop control systems are more complex to design and implement. Furthermore, measurement noise or errors in the feedback signals can lead to unexpected behavior of the system.

For dynamical systems that are subject to both disturbances and measurement noise, hybrid control systems (Fig. 11.7c) combining feedforward and feedback control can improve the total performance [19]. Hybrid control systems have frequently been applied for control tasks of the musculoskeletal system [20–22].

In model-based control, a musculoskeletal model is directly used as the controller. An inverse dynamics model can be used as a forward controller with the desired trajectories as input and optimal stimulation patterns as output. Muscle activation dynamics have to be linearized. Ferrarin et al. [21] designed a model-based feedforward control of the knee joint angle and combined it with a PID feedback controller. Jezernik et al. (2004) developed a sliding mode closed loop controller based on a musculoskeletal model for controlling the shank movement by FES.

The internal parameters of the musculoskeletal system may change because of internal disturbances such as fatigue-induced changes in muscle force generation. *Adaptive controllers* in which the controller parameters are allowed to adapt to changing plant parameters have been used to cope with such phenomena [23].

Due to their ability to map arbitrarily complex nonlinear input/output relationships from a given data set, *artificial neural networks* have been successfully applied to predict patterns of muscle stimulation needed to produce complex movements with FES-based neuroprostheses [20, 24–26]. EMG recordings from muscles under voluntary control and/or kinematic data have been used as input for the training of the neural controller [27].

11.6 Sensors

11.6.1 Artificial Sensors

Artificial sensors that are suitable for closed loop control of FES are force or pressure sensors. They are mostly placed at the point of contact, like ground contact in walking or grip force. Magnetic goniometers based on the Hall effect are used for measuring joint angles. DC accelerometers can be placed on the limb segments. Ambulatory position and orientation of human body segments can be measured accurately by combining an inertial measurement unit consisting of miniature gyroscopes, accelerometers, and magnetometers [28]. MEMS technology even allows incorporating accelerometers into injectable stimulation devices [29]. Most of the artificial sensors are placed externally on the moved limb, imposing further limitations on size, shape, and weight.

11.6.2 Natural Sensors in the Peripheral Nervous System

Natural sensors in the peripheral nervous system such as those found in the skin, muscles, tendons, and joints present an attractive alternative to artificial sensors for FES systems. Most of the peripheral sensory apparatus is still viable after injuries in the brain or spinal cord, yet not connected to the CNS.

Nerve cuff electrodes similar to the cuff electrodes for nerve stimulation have been used for chronic recording of *ENG signals* from sensory nerves. The ENG has a very small amplitude and a major source of interference is the myoelectric activity of nearby muscles. The EMG amplitude is approximately three orders of magnitude larger than the μV ENG and their spectra overlap. Implantable amplifiers have been designed, which, placed close to the recording electrode, remove EMG overlap. Control of FES thumb force using slip information obtained from the cutaneous electroneurogram has been used to control FES grip force [30].

Multicontact nerve cuff electrodes can work bidirectional by stimulating individual fascicles of nerve trunks and recording multiunit afferent activity from peripheral nerves [29].

The Utah Slanted Electrode Array USEA is inserted into the peripheral nerve for neural recording or stimulation.

11.6.3 Volitional Biological Signals

11.6.3.1 EMG

EMG is used to assess residual volitional motor activities. In so-called EMG-triggered stimulation, movement phases are initiated by volitionally activating the muscle whose EMG is measured. More sophisticated approaches establish closed loop control by modulating the stimulation intensity proportional to the measured EMG signal. EMG signals can be recorded by transcutaneous electrodes giving a noninvasive and relatively robust method for sensory input to the FES control. Any muscle the user can volitionally activate can be used for EMG recording. In the case of incomplete paralysis, it is also possible to record voluntary EMG from the same muscle that is stimulated. Bidirectional electrodes are available that can both record EMG and stimulate the muscle.

11.6.3.2 Brain Computer Interfaces

Brain computer interfaces (BCI) systems extract commands directly from the brain. The user imagines to perform a movement and brain activity signals are gained directly from the neuronal activity patterns in the corresponding motor areas of the brain. An advantage for the control of neuroprostheses is the fact that the imagined movement need not necessarily be the desired movement. Any type of command signal that is convenient for the user to generate can be used by the FES system. For example, foot movement can be imagined to trigger the FES system to open/close the hand. Due to the high inter- and intrasubject variability motor learning strategies have to be applied.

Noninvasive systems record the electroencephalogram (EEG) from the scalp or use functional magnetic resonance imaging (fMRI). The acquisition of high levels of control usually requires extensive user training. EEG-based BCIs are frequently used to trigger preprogrammed movements by FES-like hand grasp. Classifier functions are used to choose between two or more different brain states. These signals are then used like switches between different phases of a movement pattern.

11 Technical Rebuilding of Movement Function

Invasive methods use local activity from multiple neurons recorded within the brain. They show higher selectivity and are more successfully applied for complex control tasks but have the disadvantage of significant clinical risks and limited stability.

Electrocorticographic (ECoG) recording from the cortical surface has been tested as an alternative to current noninvasive and invasive recording methods [31].

First human pilot trials with both invasive and noninvasive systems suggest that BCIs could be a future option for the control of neuroprosthesis in patients with high-level SCI [32–34]. Still, there is little information available on the changes in the neural circuits in the brain after spinal cord injury, and optimal signal processing techniques have to be found to convert the existing brain signals efficiently and accurately into operative control commands.

11.7 Applications for the Lower Limb

Lower limb FES systems are used to restore walking [35], standing [36], sit-to-stand [37], cycling [38], and rowing [39]. Balance and the risk of falling are main problems in all upright body positions. As relatively high muscle forces are necessary for carrying the body weight, muscle fatigue is a serious problem in lower limb systems because it can cause falls and possible injury.

11.7.1 Cycling

Mobile FES cycling outdoors is attractive for paraplegics because they can use a standard bike (tricycle) with only a few modifications and move independently, powered by their own muscle force, over relatively long distances. Problems with balance are avoided by the seated body position, and compared to other types of movement, cycling has the advantage that the force applied to the pedal is converted into motion with very high efficiency.

FES leg cycling ergometry is frequently applied for muscle training in rehabilitation. The first commercialized leg cycling exercising system was ERGYS (Therapeutic Alliances Inc.) in 1984. So far, only external FES systems with surface electrodes have been used for FES cycling. For paraplegics, a number of leisure and sport activities are available, like basketball or hand-cycling, where only the intact upper extremities are activated. But the muscle mass of the upper extremities alone is not big enough to achieve oxygen consumption and heart rates above threshold, where the training is effective for reducing risk factors for cardiovascular and metabolic diseases. In comparison, the physiological benefits of the FES cycling training are relatively high [40].

Research on cycling by means of FES has been a focus of rehabilitation engineering at the Vienna University of Technology for several years. An instrumented

Fig. 11.8 Two-dimensional skeletal model [41]

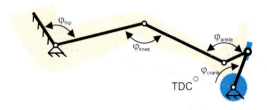

FES cycling system has been developed [38] that serves as both a stationary cycle ergometer and a mobile tricycle for paraplegics.

11.7.1.1 Simulation

A forward dynamic simulation was established to optimize the stimulation patterns for FES cycling and to determine the influence of parameter changes.

A musculoskeletal model of paraplegic isokinetic cycling on a recumbent cycle was established [41]. The skeletal model is two-dimensional and effectively consists of five rigid segments connected in frictionless hinge joints (Fig. 11.8). These segments represent the crank, foot, lower leg, upper leg, and head-arms-trunk (HAT). The point of contact between foot and pedal is under the metatarsophalangeal (MTP) joint by default. Usually, in FES cycling the ankle joint is fixed by an orthosis that also stabilizes the leg. This means that the skeletal system has only one degree of freedom and consequently the leg kinematics are entirely determined by the imposed crank kinematics, and the muscle stimulation affects the forces but does not affect the kinematics. But as the power output in FES cycling is usually quite low and overcoming the dead center is sometimes problematic, it was investigated what effect releasing the ankle joint and additionally stimulating the muscles spanning the ankle joint has on the power output and overcoming the dead center. Releasing the ankle joint adds a second degree of freedom to the linkage and thus aggravates a control problem. Not only the force applied to the pedal but also the movement has to be controlled by the muscle stimulation. On the other hand, releasing the ankle joint and additionally stimulating the muscles spanning the ankle joint brings additional physiological benefits. The equations of motion of the skeletal system were derived from the Newton–Euler equations.

The skeleton is actuated by muscles/muscle groups of the lower extremity that are stimulated during FES cycling. For fixed ankle, these are Quadriceps (Vastii and Rectus Femoris receiving identical stimulation), Gluteus Maximus, and Hamstrings. For released ankle, in addition Soleus and Gastrocnemius (receiving identical stimulation) and Tibialis Anterior are included (Fig. 11.9). A Hill-type muscle model [42] is used to represent these muscles. It consists of a contractile element, a series elastic element, and a parallel elastic element. The latter element is present in the model but has no effect in the optimal solutions. According to measurements on

11 Technical Rebuilding of Movement Function

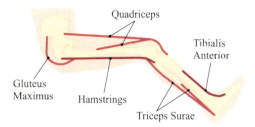

Fig. 11.9 Electrically stimulated leg muscles during FES cycling

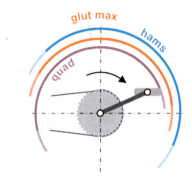

Fig. 11.10 Optimal stimulation pattern for fixed ankle joint for isokinetic FES cycling at 45 rpm. The light regions at the beginning and end of each stimulation interval indicate that the stimulation is switched on and off gradually along a ramp to avoid spasms and after-twitches

paralyzed muscles [15], the muscle activation and deactivation constants were set to 0.108 s and 0.065 s, respectively, based on the 0–70% rise time and 100–30% fall time for muscle force during isometric contraction and maximum isometric forces were set to 17% of the mean values for able-bodied subjects.

A forward dynamic simulation of isokinetic FES cycling at 30/45/60 rpm was performed. As optimization method, a parallel genetic algorithm [43] was applied. Input is muscle stimulation. The optimization criterion was to maximize mean mechanical power output over one full rotation of the crank. Figure 11.10 shows the optimal stimulation patterns for isokinetic cycling at 45 rpm, and the generated drive power is 90 W. For released ankle, the optimization results with the described model show that the ankle plantar flexors are unable to resist the torque of the pedal reaction force. To avoid this problem, it is either necessary to shorten the effective foot length by moving the position of the contact point of the foot sole and the pedal or to increase the maximal isometric force of the ankle musculature. Higher maximal isometric force of the ankle musculature might be realistic in many patients because spastic contractions reduce muscle atrophy. Shortening the effective foot length from 0.165 to 0.055 m resulted in a 10% power increase. Double maximal isometric force in the ankle musculature plus shortening the effective foot length

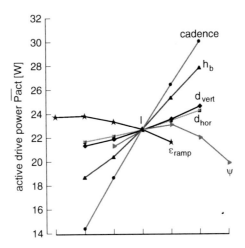

Fig. 11.11 Active power output of all stimulated muscles summed up and influence of variation of the single parameters. All curves pass through the point I where the optimization is performed with an initial average set of parameters (details in [44]). h_b .. body height; d_{vert}, d_{hor} ... vertical and horizontal distances between crank axis and hip joint, respectively; ψ ... backrest angle (from vertical), ε_{ramp} ... length of ramp at begin and end of stimulation

to 0.11 m shows a 29% power increase and the generated drive torque is positive over the full crank rotation what means that overcoming the dead center might be facilitated.

FES-cycling performance is influenced by a number of parameters such as seating position, physiological parameters, conditions of surface stimulation, and pedaling rate. A sensitivity analysis was performed to determine the influence of the most important parameters on optimal muscle stimulation patterns and power output of FES cycling (Fig. 11.11) [44].

The results of the simulation show in which regions the individual leg muscles should be stimulated and what influence some parameters have on stimulation and power output. Still there is a number of parameters which cannot be considered adequately in the simulation. These include unpredictable spasm activity which also strongly depends on daily condition of the patient, co-stimulation of antagonists, muscle fatigue, and muscle force hysteresis, as the muscle force depends on both activation and movement history. Also muscle condition is very different among patients due to factors such as training status, spasms, type of and time since injury. It is important to test the relation between stimulation parameters and generated muscle forces for each patient individually.

11.7.1.2 Instrumented FES Cycling System

For details see [38].

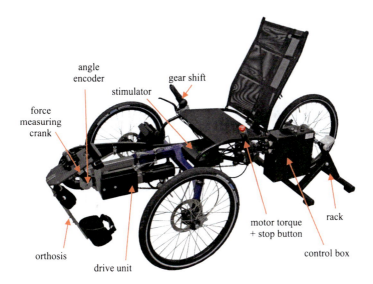

Fig. 11.12 Instrumented test- and training system and main components. The control box contains the motor control and the accumulators

Mechanical Design

A commercially available tricycle was adapted as the basic frame for the FES-cycling system (see Fig. 11.12). The horizontal distance between the crank bearing and the seat can easily be adjusted to different leg lengths. For easier transfer from a wheelchair to the tricycle, the right steering handle can be dismounted by opening a quick clamp and a transfer board hugged on to the frame.

Orthoses

Figure 11.13 shows the orthoses which are mounted on the pedals. Their function is to stabilize the legs in the parasagittal plane during pedaling. Due to the telescopic shaft, the orthoses are adaptable to the length of the users shank.

Force Measurement Cranks

The force measurement cranks are based on strain gauge technology. Strain gauges are arranged in three full Wheatstone bridges on the aluminum corpus of the cranks. The arrangement of the strain gauges allows measurement of the radial force (in the direction of the crank), the tangential force (rectangular to the crank in the pedaling plane), and the torque around the longitudinal axis of the crank. The signals of each crank are amplified, digitized, and sent to a laptop computer. A specialized time

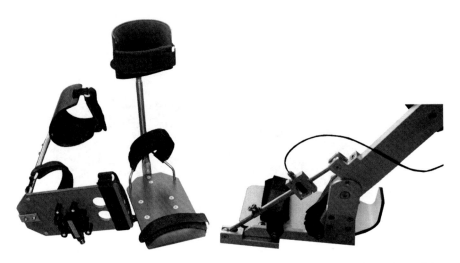

Fig. 11.13 Orthoses for leg stabilization in the parasagittal plane. *Left*: lightweight orthoses fixing the ankle joint; *right*: orthosis with a ball bearing and adjustable movement range at the ankle and a force measuring unit for measurements on the generated ankle torque

synchronized telemetry network is used to ensure time-correlated measurements of left and right leg.

Electrical Stimulation

Figure 11.14 shows the current-controlled 10-channel stimulator that was developed for FES cycling and rowing applications. The stimulator induces biphasic rectangular pulses with a stimulation current from 0 to 150 mA (at 1 kΩ), frequency from 0 to 100 Hz, and pulse width from 0 to 700 µs. In "time mode" the stimulation is delivered as a function of time, in "angle mode" as a function of an angle signal that is processed in the stimulator. The stimulator automatically shifts the stimulation pattern backward as a function of the actual cadence to consider the dynamic characteristics (activation and deactivation times) of the muscles. Up to three sets of individual stimulation patterns can be stored in the stimulator. During FES cycling, the preset stimulation currents can be scaled from 0 to 100% by turning an adjusting knob on the right steering bar of the FES cycle.

Drive Train

The main component of the drive train is the motor unit consisting of a servo-motor and a planetary gear. An electromagnetic coupling connects the motor unit to a bevel gear on which the pinion is mounted. The pinion is then connected to the cranks by a chain.

11 Technical Rebuilding of Movement Function

Fig. 11.14 Current controlled 10-channel stimulator for FES cycling and rowing applications

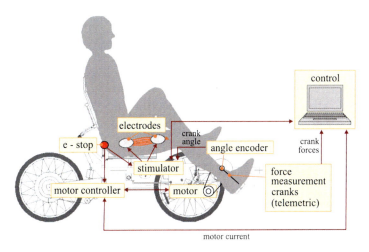

Fig. 11.15 Schematic of a FES cycling system with advanced instrumentation and control for determination of individual optimal stimulation patterns

Control

A schematic of the control of the FES cycling system is shown in Fig. 11.15. Basically, two operation modes are available:

- Mobile cycling

The electrical stimulation of the muscles/muscle groups Quadriceps, Hamstrings, and Gluteus Maximus of each leg is generated by the 10-channel current-controlled stimulator via attached surface electrodes. The stimulation patterns, i.e., the crank angle intervals where each muscle/muscle group should be stimulated for maximum power output, are stored in the stimulator. The angular position of the crank is detected by an angle encoder which is integrated in the crank axis and read into the stimulator.

The motor generates a constant drive torque which is controlled by a turning handle on the left steering bar of the cycle by the user. If no motor support is necessary, the motor can be decoupled by switching off the electromagnetic coupling.

The gears of the gear hub in the back wheel are changed by turning the outer part of the right steering bar.

- Stationary cycling and measurement mode

For stationary cycling, the cycle is hooked up on a rack.

Enhanced control for both stimulator and motor is available by connecting a computer to the system via an RS 232 Interface. A LabView-based control program then communicates with the stimulator and the motor control, and reads in motor current and crank angle data and additionally force data from the force measuring cranks. The control program offers two modes of operation: an expert mode with unlimited access to all parameters of the muscle stimulation and motor control, and a wizard mode with limited access which guides the user step by step through a predefined series of training and measurement units. The data processing is automated and the measurement data of all the training units of one person are stored together with the personal log file.

To allow reproducible force measurements, the motor can either move the cranks at constant angular velocity for isokinetic measurements or hold the cranks on a defined position for isometric measurements. Predefined measurement routines allow to determine individual sets of optimal stimulation parameters based on the results of the mathematical simulation. At first, isometric measurements determine the muscle's force response and adequate stimulation intensity for cycling, then the crank angle interval is determined, in which the muscle applies positive crank torque. Isokinetic measurements then adapt the optimal stimulation interval to higher cadences.

11.7.1.3 Clinical Application

The FES cycling system offers multifunctional equipment for FES-cycling training and therapy. Currently, the system is being tested as a rehabilitation tool in clinical rehabilitation for spinal cord-injured subjects in a clinical study in cooperation with the AUVA Rehabilitation Center Weisser Hof in Austria. Patients are doing an FES training session three times a week over 2 months as part of their rehabilitation program and also do outdoor cycling. Figure 11.16 shows a paraplegic subject performing stationary training on the FES tricycle. Due to the force measurements and the automated data processing, the therapy progress can be well monitored. Also spasticity is assessed before and after each FES session, and it has been shown in accordance with earlier studies that the FES training reduces spasticity at least temporarily [45].

Fig. 11.16 Paraplegic subject performing stationary training on the FES tricycle

11.7.2 Rowing

Ergometer rowing with FES of the muscles in the lower extremity enables paraplegics to participate in this mainstream activity for health, leisure, and sport. As muscle mass of both upper and lower extremities is metabolically active during the rowing motion, the cardiovascular training is higher than in exercises where only the muscle mass of the lower extremities is activated (Hettinga et al. 2004).

11.7.2.1 Simulation

For determination of the optimal stimulation patterns for ergometer rowing by means of FES, a musculoskeletal model was established in Matlab Simulink (Kuchler and Gföhler, 2004). The model of the user-rowing machine system represents a planar 8-link kinematic chain with three degrees of freedom (Fig. 11.17). The resistance mechanism was modeled by Euler's principal equation for flow machinery (damping element k_L in Fig. 11.17). The equations of motion were derived using the Newton–Euler equations.

Seventeen muscle groups of the upper and lower extremities were considered. The muscles of the lower extremities are activated by surface electrodes. Maximum isometric forces were scaled according to measurements [15]. The three degrees of freedom represented by the angles φ_1, φ_3, and φ_6 and the vertical contact force at the seat were used as inputs to solve the inverse dynamic problem [46]. The muscle forces were determined using mathematical optimization [16]. Figure 11.18 shows the normalized muscle forces in the lower extremities over one complete rowing cycle. The results show that a high muscle force from Iliopsoas is necessary for hip flexion at the beginning of the recovery phase. But the deep-lying Iliopsoas muscle

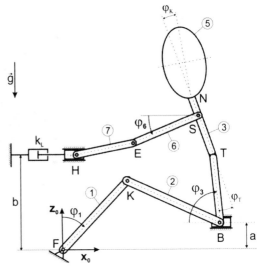

Fig. 11.17 Model of the paraplegic-rowing machine system

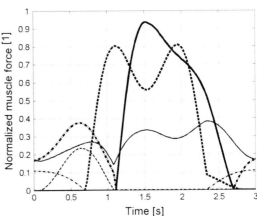

Fig. 11.18 Forces of Tibialis anterior (*thick dashed*), Soleus and Gastrocnemius (*thin dash-dot*), Vastii and Rectus femoris (*thin dashed*), Hamstrings (*thin solid*), Gluteus maximus (*thick dash-dot*), and Iliopsoas (*thick solid*) normalized by the corresponding maximum isometric force during one full rowing stroke

cannot be stimulated with surface electrodes; consequently, there will be difficulties with generating a sufficient hip flexion torque, and this has to be considered in ergometer design.

11.7.2.2 Instrumented FES Rowing Ergometer

For experimental investigations on FES rowing, an instrumented rowing ergometer was designed based on a standard Concept II (Concept2 Deutschland GmbH, Hamburg, DE) ergometer [47]. The sliding seat was replaced by a construction with soft seating and adjustable backrest for stability of the upper body. Orthoses are fixed on the foot rest for side-to-side stability of the legs. Five muscles/muscle

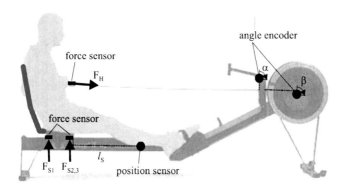

Fig. 11.19 Instrumented FES rowing ergometer

groups are stimulated in each leg by surface electrodes: Tibialis anterior, Soleus and Gastrocnemius, Vastii and Rectus femoris, Hamstrings, and Gluteus maximus.

Instrumentation of the ergometer allows the determination of actual kinematics and kinetics for further investigations and individual optimization of the rowing motion. The vertical seat force is measured by a combination of three force measuring cells so that the torques can be eliminated. A force measuring cell which is placed between handle and chain gives the pull force at the handle. The horizontal position of the seat is measured by a position sensor. Two angle decoders are used for measuring the length and angular position of the chain from the handle to the resistance mechanism and hence define the two-dimensional position of the handle. The digital signals from the angle decoders and the position sensor are read in the control via a microprocessor. The analog signals from the force measuring cells are read into a data acquisition card and from there also to the control. The joint trajectories of all three degrees of freedom can be calculated from the measured position data. Together with the measured vertical seat force, all necessary input data are available to solve the inverse dynamic problem with subject-specific input data (Fig. 11.19).

The control for the FES rowing motion is based on the same LabView program as for FES cycling; the 10-channel stimulator which was developed for FES cycling was extended with a rowing mode where the stimulation of the leg muscles is controlled by the horizontal position of the seat which is detected by a position sensor. The range of horizontal movement of both handle and seat are defined as a function of the leg length for each individual subject and the stimulation pattern is scaled accordingly.

11.7.3 Gait

Foot drop is a gait-limiting factor in patients with stroke or other disorders of the CNS. Due to weakness of the ankle dorsiflexors, the foot drops and may drag on the ground during the swing phase of gait. The first system to correct foot drop was proposed in 1961 by Liberson. In this device, the common peroneal nerve was stimulated when the heel came off the ground to flex the ankle and thereby lift the foot. A few systems based on this principle are commercially available. For the stimulation either surface electrodes on the anterior leg muscles or implanted electrodes near the peroneal nerve are used, the stimulation is synchronized to the gait phase by either a switch under the heel or neural signals recorded from the sural nerve with a cuff electrode or inertia sensors on the foot (WalkAide®, Innovative Neurotronics; Veltnik 2003; Weber et al. 2005).

Walking by means of FES is an attractive and desirable goal for many patients because it gives them the ability to move as people do naturally. However, mainly due to the problem of keeping balance and weight carrying, equipment for FES walking is cumbersome to don and operate and only safe to operate under highly controlled conditions. During FES walking, patients usually have to use a rollator or crutches what makes it less applicable in daily life. The first systems enabling walking were external devices that were developed by Kralj et al. [48]. Currently mainly systems with surface electrodes are used, but also fully implanted systems with stimulation of up to 48 muscles are available. The implantation frees the patient from cabling, but still a walking frame or crutches are necessary. All available systems are open loop controlled. The user has to initiate the step, and then preprogrammed stimulation sequences are carried out for the whole step cycle. Hybrid systems combine electrical stimulation with mechanical bracing and can potentially combine the best features of mechanical bracing and FES into new systems for walking after SCI that offer more advantages than the individual components acting alone.

11.8 Applications for the Upper Limb

FES systems restoring upper limb function are mainly suitable for tetraplegics with cervical lesions; the peripheral nerves innervating the arm muscles must be intact. There are systems based on surface electrodes and percutaneous electrodes as well as fully implanted systems available. The higher the level of the spinal cord injury, the more joints must be controlled. Also the higher the injury, the fewer biological control signals are available. FES has so far concentrated on restoring grasp in patients with C5/C6 injuries. If the injury occurs below this level, tendon transfer of functioning muscles is used to recreate the ability to grasp objects. At injury levels above C4, it is difficult to stabilize the arm. Shoulder and elbow usually retain some voluntary function in C5/6 spinal cord-injured people, but usually the arm is weak due to the paralysis of some key muscles.

The Bionic Glove [49] is a completely noninvasive FES device designed to activate hand muscles in C6/C7 spinal cord-injured persons who have some active wrist movement. It consists of self-adhesive surface electrodes which are placed above the motor points of the muscles to be stimulated and a glove which contains a wrist position sensor and a box containing stimulator and control. Stimulation of the muscles that produce hand grasp is triggered by extending the wrist. Stimulation of the muscles that produce hand opening is triggered by flexing the wrist.

A clinically accepted example for an upper extremity neuroprosthesis for persons with C5/C6 spinal cord injury is the Freehand System [50] which consists of fully implanted electrodes and stimulator and an external control unit including an inductive coil and a shoulder position sensor: Eight epimysial electrodes control palmar and lateral hand grasp by neuromuscular stimulation of hand and arm muscles. Hand grasp is triggered through operation of an external joystick, controlled by the movement of the opposing nonparalyzed shoulder, which through a radiofrequency-powered and -controlled implanted stimulator delivers electrical stimulation. The Freehand system gives hundreds of persons the ability to feed and groom themselves; some can even operate a computer with their hand. In order to establish closed loop control of the Freehand system, nerve cuff electrodes have been used to record activity from cutaneous mechanoreceptors at the index finger to determine grip force.

11.9 Outlook

FES-based neuroprostheses for upper and lower extremities evoke lost movement functions in SCI subjects. Exercises such as FES cycling and rowing enable paraplegics to participate in mainstream activities and improve their health and fitness by exercising like able-bodied subjects. The completely external systems are relatively easy to don and handle, but the possibilities for control and selective muscle actuation are limited. Besides the physiological training, a main desire of the patients is to be able to move naturally as able-bodied subjects do. For use in daily living the neuroprosthesis should support the user in a cooperative way. It should respond to the moment-to-moment needs of the user without unduly distracting his attention. The user should be released of tasks that can be automatized but keep control over the movement coordination.

Recent developments in technology allow the design of miniaturized devices with sophisticated control systems. But still it is a challenge to detect what movement the patient desires to make and to transfer the signal proportionally into muscle force.

One crucial point is signal processing of biosignals that are recorded from brain and muscle activity to get a reliable and reproducible proportional command signal.

Currently, another limiting point are the electrodes because the correlation between stimulation and muscle force is difficult to predict. One major problem here is also fatigue which is a main issue in all FES applications.

At present, implanted systems use either batteries, disposable or rechargeable, or external AC power as power supply. Currently it is investigated whether biothermal power sources which use small temperature gradients in the body to create electrical power could make body powered implants possible (Biophan Technologies of West Henrietta).

Today the majority of spinal cord-injured subjects has an incomplete lesion. Consequently, there is a need for modular programmable systems that can be adapted to individual situations. FES with surface electrodes can be problematic for subjects with incomplete SCI, because when the sensory function is still intact the electrical field can evoke sensations of pain already at low stimulation intensities where only low muscle forces are generated.

More sophisticated systems will be available in the future, both external and implanted. Still besides technology also other personal factors have to be considered in the decision if a system is appropriate for an individual SCI subject or not. From a social point of view, patients should certainly not be dependent on using a neuroprosthesis for daily living. But each patient should have the freedom to choose using a neuroprosthesis for more independent daily living and self-responsible exercising for health and fitness.

Even if a cure of spinal cord injury should be possible in the future, those patients who have exercised their body and kept their musculoskeletal structures in shape will benefit first and most of it.

References

1. W.T. Liberson, H.J. Holmquest, D. Scot, M. Dow, Stimulation of the peroneal nerve, synchronized with the swing phase of the gait of hemiplegic patients. Arch. Phys. Med. Rehabil. **42**, 101–105 (1961)
2. T.W.J. Janssen, R.M. Glaser, D.B. Shuster, Clinical efficacy of electrical stimulation exercise training: effects on health, fitness, and function. Top. Spinal Cord Inj. Rehabil. **3**, 33–49 (1998)
3. S. Malagodi, M.W. Ferguson-Pell, R.D. Masiello, A functional electrical stimulation exercise system designed to increase bone density in spinal cord injured individuals. IEEE Trans. Rehab. Eng. TRE **1**, 213–219 (1993)
4. F. Rattay, Modeling the excitation of fibers under surface electrodes. IEEE Trans. Biomed. Eng. BME **35**, 199–202 (1988)
5. M. Solomonow, External control of the neuromuscular system. IEEE Trans. Biomed. Eng. BME **31**, 752–763 (1984)
6. W. Reichenfelser, M. Gföhler, T. Kakebeeke, P. Lugner, G. Feik, Determination of efficient stimulation patterns for FES – cycling; IFMBE Proc. EMBEC 05 Prague (2005)
7. W.K. Durfee, K.I. Palmer, Estimation of force-activation, force-length, and force-velocity properties in isolated, electrically stimulated muscle. IEEE Trans. Biomed. Eng. BME **41**, 205–216 (1994)
8. M. Lawrence, G. Gross, M. Lang, A. Kuhn, T. Keller, M. Morari et al., Assessment of finger forces and wrist torques for functional grasp using new multichannel textile neuroprostheses. Artif. Organs **32**(8), 634–638 (2008)
9. T.G. McNaughton, K.W. Horch, Metallized polymer fibers as leadwires and intrafascicular microelectrodes. J. Neurosci. Methods **70**(1), 103–110 (1996)
10. A. Branner, R.B. Stein, R.A. Normann, Selective stimulation of cat sciatic nerve using an array of varying-length microelectrodes. J. Neurophysiol. **85**(4), 1585–1594 (2001)
11. N. Lago, K. Yoshida, K.P. Koch, X. Navarro, Assessment of biocompatibility of chronically implanted polyimide and platinum intrafascicular electrodes. IEEE Trans. Biomed. Eng. **54**(2), 281–290 (2007)
12. G.E. Loeb, R. Peck, W.H. Moore, K. Hood, BION system for distributed neural prosthetic interfaces. Med. Eng. Phys. **23**(1), 9–18 (2001)

13. J.M. Winters and S.L.-Y. Woo(eds), *Multiple Muscle Systems* (Springer Verlag, New York, 1990)
14. F.E. Zajac, Muscle and tendon: properties, models, scaling, and application to biomechanics and motor control. Crit. Rev. Biomed. Eng. **17**(4), 359–411 (1989) (Review)
15. M. Gföhler, J. Wassermann, P. Eser, T. Kakebeeke, H. Lechner, W. Reichenfelser et al., Muscle behavior in artificially activated muscle – measurements on neurologically intact and paraplegic subjects, in *International Society of Biomechanics XIXth Congress*, ed. by P. Milburn, B. Wilson, T. Yanai (International Society of Biomechanics, Dunedin, 2003), p. 5
16. R.D. Crowninshield, R.A. Brand, A physiologically based criterion of muscle force prediction in locomotion. J. Biomech. **14**(11), 793–801 (1981)
17. M.G. Pandy, F.E. Zajac, E. Sim, W.S. Levine, An optimal control model for maximum-height human jumping. J. Biomech. **23**(12), 1185–1198 (1990)
18. A. Seth, M.G. Pandy, A neuromusculoskeletal tracking method for estimating individual muscle forces in human movement. J. Biomech. **40**(2), 356–366 (2007)
19. A.D. Kuo, The relative roles of feedforward and feedback in the control of rhythmic movements. Motor Control **6**(2), 129–145 (2002)
20. G.C. Chang, J.J. Luh, G.D. Liao, J.S. Lai, C.K. Cheng, B.L. Kuo et al., A neuro-control system for the knee joint position control with quadriceps stimulation. IEEE Trans. Rehab. Eng. **5**(1), 2–11 (1997)
21. M. Ferrarin, F. Palazzo, R. Riener, J. Quintern, Model-based control of FES-induced single joint movements. IEEE Trans. Neural Syst. Rehabil. Eng. TRE **9**(3), 245–257 (2001)
22. H. Park, D.M. Durand, Motion control of musculoskeletal systems with redundancy. Biol. Cybern. **99**(6), 503–516 (2008)
23. L.A. Bernotas, P.E. Crago, H.J. Chizeck, Adaptive control of electrically stimulated muscle. IEEE Trans. Biomed. Eng. **BME-34**(2), 140–147 (1987)
24. M. Goffredo, I. Bernabucci, M. Schmid, S. Conforto, A neural tracking and motor control approach to improve rehabilitation of upper limb movements. J. Neuroeng. Rehabil. **5**, 5 (2008)
25. L. Johnson, A.J. Fuglevand, Evaluation of probabilistic methods to predict muscle activity: implications for neuroprosthetics. J. Neural Eng. **6**(5), 55008 (2009)
26. J.L. Lujan, Crago PE automated optimal coordination of multiple-DOF neuromuscular actions in feedforward neuroprostheses. IEEE Trans. Biomed. Eng. BME **56**(1), 179–187 (2009)
27. J.J.G. Hincapie, R.F. Kirsch, Feasibility of EMG-based neural network controller for an upper extremity neuroprosthesis. IEEE Trans. Neural Syst. Rehabil. Eng. **17**(1), 80–90 (2009)
28. D. Roetenberg, P.J. Slycke, P.H. Veltink, Ambulatory position and orientation tracking fusing magnetic and inertial sensing. IEEE Trans. Biomed. Eng. **54**(5), 883–890 (2007)
29. G.E. Loeb, R. Davoodi, The functional reanimation of paralyzed limbs. IEEE Eng. Med. Biol. Mag. **24**(5), 45–51 (2005)
30. M. Haugland, A. Lickel, J. Haase, T. Sinkjaer, Control of FES thumb force using slip information obtained from the cutaneous electroneurogram in quadriplegic man. IEEE Trans. Rehabil. Eng. **7**(2), 215–227 (1999)
31. K.J. Miller, M. denNijs, P. Shenoy, J.W. Miller, R.P. Rao, J.G. Ojemann et al., Real-time functional brain mapping using electrocorticography. Neuroimage **37**(2), 504–507 (2007)
32. G.R. Müller-Putz, R. Scherer, G. Pfurtscheller, R. Rupp, EEG-based neuroprosthesis control: a step towards clinical practice. Neurosci. Lett. **382**(1–2), 169–174 (2005)
33. G. Pfurtscheller, G.R. Müller, J. Pfurtscheller, H.J. Gerner, R. Rupp, 'Thought'–control of functional electrical stimulation to restore hand grasp in a patient with tetraplegia. Neurosci. Lett. **351**(1), 33–36 (2003)
34. Y. Song, D. Borton, S. Park, W.R. Patterson, C.W. Bull, F. Laiwalla et al., Active microelectronic neurosensor arrays for implantable brain communication interfaces. IEEE Trans. Neural Syst. Rehabil. Eng. **17**(4), 339–345 (2009)
35. T.A. Thrasher, M.R. Popovic, Functional electrical stimulation of walking: function, exercise and rehabilitation. Ann. Readapt. Med. Phys. **51**(6), 452–460 (2008) (Epub 2008 Jun 18. Review. English, French)

36. G.P. Braz, M. Russold, R.M. Smith, G.M. Davis, Efficacy and stability performance of traditional versus motion sensor-assisted strategies for FES standing. J. Biomech. **42**(9), 1332–1338 (2009)
37. R. Davoodi, B.J. Andrews, Optimal control of FES-assisted standing up in paraplegia using genetic algorithms. Med. Eng. Phys. **21**(9), 609–617 (1999)
38. W. Reichenfelser, H. Hackl, S. Mina, S. Hanke, P. Lugner, M. Gföhler, Trainings- and measurement-system for FES-cycling, in *"IFESS 2008- from movement to mind"*, Biomedizinische Technik, vol. 53, Suppl.1 (Berlin, New York, 2008), pp. 265–267. ISSN: 0939–4990
39. R. Davoodi, B.J. Andrews, G.D. Wheeler, R. Lederer, Development of an indoor rowing machine with manual FES controller for total body exercise in paraplegia. IEEE Trans. Neural Syst. Rehabil. Eng. TRE **10**(3), 197–203 (2002)
40. F. Ché, G.M. Davis, Cardiovascular and metabolic responses during functional electric stimulation cycling at different cadences. Arch. Phys. Med. Rehab. **89**(4), 719–725 (2008)
41. A.J.K. van Soest, M. Gföhler, R. Casius, Consequences of ankle joint fixation on FES cycling power output; a simulation study. Med. Sci. Sport. Exercise **1**, S.797–S.806 (2005)
42. A.V. Hill, The heat of shortening and dynamics constants of muscles. Proc. R. Soc. Lond. B (London: Royal Society) **126**(843): 136-195 (October 1938)
43. A.J. van Soest, L.J.R. Casius, The merits of a parallel genetic algorithm in solving hard optimization problems. J. Biomech. Eng. **125**:141–146 (2003).
44. M. Gföhler, P. Lugner, Dynamic simulation of FES-cycling: influence of individual parameters. IEEE Trans. Neural Syst. Rehabil. Eng. **12**(4), 398–405 (2004)
45. W. Reichenfelser, H. Hackl, J. Wiedner, J. Hufgard, K. Gstaltner, S. Mina, S. Hanke, M. Gföhler: Influence of FES cycling on spasticity in subjects with incomplete spastic paraplegia; in *Rehabilitation: Mobility, Exercise & Sports*, Assistive Technology Research Series, vol 26 (IOS Press, Amsterdam, 2010), pp. 317–319. ISBN: 978-1-60750-080-3
46. M. Kuchler, M. Gföhler, Development of a biomechanical model of the human body including the upper and the lower extremities used to simulate the motion in a rowing ergometer – the inverse dynamic problem, in *Congress Handbook and Book of Abstracts of the International Society of Biomechanics XIXth Congress*, 4 pages (2003), p. 122
47. I. Hauer, Instrumentierter Ruderergometer für Anwendung in Sport und Rehabilitation, Betreuer/in(nen), Konstruktionswissenschaften und Technische Logistik, ed. by M. Gföhler (2007) (Diploma Thesis, May 2007)
48. A.R. Kralj, T. Bajd, M. Munih, R. Turk, FES gait restoration and balance control in spinal cord-injured patients. Prog. Brain. Res. **97**, 387–396 (1993)
49. A. Prochazka, M. Gauthier, M. Wieler, Z. Kenwell, The bionic glove: an electrical stimulator garment that provides controlled grasp and hand opening in quadriplegia. Arch. Phys. Med. Rehabil. **78**, 608–614 (1997)
50. J.J. Pancrazio, P.H. Peckham, Neuroprosthetic devices: how far are we from recovering movement in paralyzed patients? Expert Rev. Neurother. **9**(4), 427–430 (2009)
51. M. Kuchler, M. Gföhler, Mechanical modeling and simulation of the human movement on a rowing ergometer. Simul. News Eur. **40**, S.10–S.19 (2004). ISSN 0929–2268
52. P.D. Faghri, R.M. Glaser, S.F. Figoni, Functional electrical stimulation leg cycle ergometer exercise: training effects on cardiorespiratory responses of spinal cord injured subjects at rest and during submaximal exercise. Arch. Phys. Med. Rehabil. **73**(11), 1085–1093 (1992)
53. T. Mohr, J.L. Andersen, F. Biering-Sørensen, H. Galbo, J. Bangsbo, A. Wagner, M. Kjaer, Long-term adaptation to electrically induced cycle training in severe spinal cord injured individuals. Spinal Cord **35**(1), 1–16 (1997)
54. S. Jezernik, R.G. Wassink, T. Keller, Sliding mode closed-loop control of FES: controlling the shank movement. IEEE Trans. Biomed. Eng. **51**(2), 263–272 (2004)
55. D.M. Hettinga, B.J. Andrews, G.D. Wheeler, J.Y. Jeon, J. Verellen, J.J. Laskin, L.M. Olenik, R. Lederer, R. Burnham, R.D. Steadward, FES-rowing for persons with spinal cord injury, in *Proceedings of the 9th Annual Conference of the International FES Society September 2004*, Bournemouth, 2004

56. P.H. Veltink, P. Slycke, J. Hemssems, R. Buschman, G. Bultstra, H. Hermens, Three dimensional inertial sensing of foot movements for automatic tuning of a two-channel implantable drop-foot stimulator. Med. Eng. Phys. **25**(1), 21–28 (2003)
57. D.J. Weber, R.B. Stein, K.M. Chan, G. Loeb, F. Richmond, R. Rolf, K. James, S.L. Chong, BIONic WalkAide for correcting foot drop. IEEE Trans. Neural Syst. Rehabil. Eng. **13**(2), 242–246 (2005)

Chapter 12
Improving Hearing Performance Using Natural Auditory Coding Strategies

Frank Rattay

Abstract Sound transfer from the human ear to the brain is based on three quite different neural coding principles when the continuous temporal auditory source signal is sent as binary code in excellent quality via 30,000 nerve fibers per ear. Cochlear implants are well-accepted neural prostheses for people with sensory hearing loss, but currently the devices are inspired only by the tonotopic principle. According to this principle, every sound frequency is mapped to a specific place along the cochlea. By electrical stimulation, the frequency content of the acoustic signal is distributed via few contacts of the prosthesis to corresponding places and generates spikes there. In contrast to the natural situation, the artificially evoked information content in the auditory nerve is quite poor, especially because the richness of the temporal fine structure of the neural pattern is replaced by a firing pattern that is strongly synchronized with an artificial cycle duration. Improvement in hearing performance is expected by involving more of the ingenious strategies developed during evolution.

12.1 The Hair Cell Transforms Mechanical into Neural Signals

The exceptional performance and the extreme high sensitivity of the auditory system are excellent examples of evolution. It was developed together with the lateral line organ of fishes, a sensory organ that consists of a canal running along both sides of the body, communicating via sensory pores through scales to the exterior. Analyzing the vibrations of the surrounding water as spatial and temporal functions, the lateral line system helps the fish to avoid collisions, to orient itself in relation to water currents, and to locate prey. This way it is a touch sense over distance.

F. Rattay (✉)
Institute for Analysis and Scientific Computing, Vienna Technical University, Vienna, Austria
e-mail: frank.rattay@tuwien.ac.at

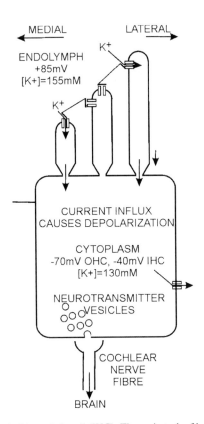

Fig. 12.1 Scheme of a typical inner hair cell (IHC). The main task of hair cells in sensory systems is to detect forces which deflect their hairs (stereocilia). The apical part of the cell including the stereocilia enters the endolymphatic fluid, which is characterized by its high potassium concentration [K^+]. The transmembrane voltage of -40 mV for IHC and -70 mV for outer hair cells (OHC) is caused by the K concentration gradient between cell body and cortilymph. Current influx that changes the receptor potential occurs mainly through the ion channels of the stereocilia: stereociliary displacement to the lateral side of the cochlea causes an increase in ion channel open probability and hence depolarization of the receptor potential, whereas stereociliary displacement to the medial side results in hyperpolarization. Small variations in the receptor potential (about 0.1 mV) cause a release of neurotransmitter which may cause spiking in the most sensitive fibers of the auditory nerve [1]

Among such amphibians as frogs, lateral line organs and their neural connections disappear during the metamorphosis of tadpoles because as adults they need no longer to feed under water. The higher land-inhabiting vertebrates (reptiles, birds, and mammals) do not possess the lateral line organs anymore. However, their deeply situated labyrinthine sense organs use the same principle of detecting information from fluid motion via hair cells (Fig. 12.1). Cell membrane voltage fluctuations of a single hair cell are able to respond synchronized to high audible input frequencies of 20 kHz in man, and to even essentially higher frequencies in bats, whales, or dolphins.

12 Improving Hearing Performance Using Natural Auditory Coding Strategies

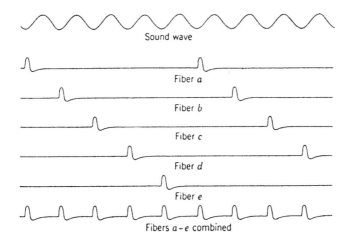

Fig. 12.2 The volley principle. A periodic acoustic stimulus will cause firing patterns in different axons innervating one inner hair cell, e.g., as marked in this regular example with (a)–(e). The combined signal contains all the minima of sound even when no single fiber is able to fire with the high frequency of the source signal. After [2]

Detailed temporal information resulting from the mechanical input signal is recorded by the hair cell as intracellular potential fluctuations. But most of the temporal fine structures would be lost if a single neural connection as schematically shown in Fig. 12.1 has to handle signal transfer to the brain because of the slower operation of neural signals. This problem is solved by a change from serial to parallel coding: a single human inner hair cell has in average eight synaptic release zones that distribute the information to eight spiking nerve fibers. This method is called volley principle (Fig. 12.2).

In the following, we will introduce the three differing neural auditory coding principles used by nature to handle the wide range of 120 dB for audible signal amplitudes with hair cells of extremely smaller operating ranges. With cochlear implants, many deaf people obtain auditory perception by electrical stimulation of the auditory nerve. A disadvantage in hearing quality with the currently available devices is, however, that signal processing strategies mimic only a single method of the three natural principles.

12.2 The Human Ear

In the healthy ear, the acoustic signal which is physically a vibration of air pressure causes an analogous movement of the hairs of the hair cells because the motion is mechanically transformed into the liquid environment (Fig. 12.3). The hair cells finally transform the acoustical input into neural signals. Frequency and loudness

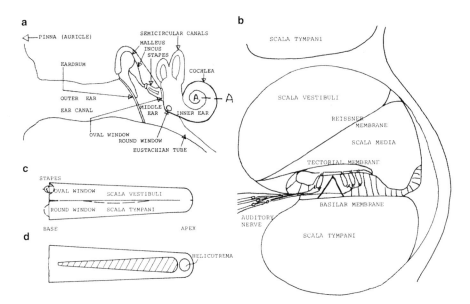

Fig. 12.3 (a) The main parts of the human ear. Part of a cross section marked as A–A is shown in (b). (c) and (d) are front and top view schemata of the uncoiled cochlea. The shaded area in (d) represents the basilar membrane which is the elastic part between the upper and the lower chambers. The 35-mm long basilar membrane is stiffer against the basal end but broadens from 0.1 to 0.5 mm to the apical end and it changes the resonance properties continuously. When a tone starts with a compression, the stapes will press the oval window membrane inside (*broken lines* in (c)). As a consequence, part of the perilymphatic fluid will escape through the helicotrema, which connects the scala vestibule and the scala tympani. The higher pressure in the scala vestibuli bends also the basilar membrane downward. The broad widening (*long broken line* in (c)) sharpens with time when sound is presented periodically at constant frequency. (b) If the pressure in the scala vestibuli is higher than that in the scala tympani, the organ of Corti and other parts of the scala media move down due to the compliance of the basilar membrane. The resistance of Reisner's membrane is very low and can be neglected. The movement of the basilar membrane will be registered by the organ of Corti, which transforms the mechanical movement of the basilar membrane into neural signals with the help of the inner hair cells. At the base of an inner hair cell, several synapses are situated. Each of these synapses is connected with an axon of the auditory nerve. The axons of the inner hair cells send their information to the brain (afferent neurons). Three rows of outer hair cells receive their information mostly from the brain (efferent neurons)

are the two essential items of the acoustical input that are transmitted via nerve fibers from the inner ear to the brain. Sometimes the cochlea is called a frequency analyzer because of its mechanical tuning properties. This way frequency sensitivity is a function of the place of stimulation, i.e., the hair cells at the beginning of the cochlea response sensitively to high frequencies, whereas only deep tones will reach the hair cells at the apical end of the hearing organ. Increasing the sound intensity causes higher firing rates in an increasing number of auditory nerve fibers.

12.3 Place Theory Versus Temporal Theory

There is an old controversy whether the inner ear uses the place coding principle or the temporal structure of the spiking pattern to transmit the information of an acoustic signal. The "place theory" was introduced by Helmholtz. He assumed already that: (1) the basilar membrane consists of segments with varying stiffness, (2) frequency selectivity is based on the resonance of a corresponding part, and (3) each part is connected with a nerve fiber [3]. The most extreme form of the "temporal theory" was put forward by Rutherford, who speculated that each hair cell in the cochlea responds to every tone entering the inner ear [4].

Hundred years later both theories were applied independently as basis for two essentially different signal processing strategies in cochlear implants, and every success concerning acoustic perception seemed to underline the validity of just one of both theories. Early implants transmitted the electrical stimulus as a continuous voltage, analog to an amplitude compressed acoustical input to a single active electrode placed, e.g., in the scala tympani close to the nerve fibers of the auditory nerve [5]. As vowel discrimination is possible with such a single channel method, this was a proof against a pure place theory. On the other hand, all modern cochlear implants distribute the signals based only on the frequency distribution in the acoustic source signal according to the tonotopic organization of the cochlea that is according to the place principle. Neglecting the temporal fine structure of the input is one reason for deficient speech understanding with cochlear implants, especially in noisy environment. In a computer simulation study of inner ear mechanics, we demonstrated that the sharp frequency location needs time for development [6] and it may be more important for music than speech perception responses.

We should learn from hearing physiology that the brain uses several strategies in parallel to obtain the relevant information from the auditory input signal. One method which is not much considered is the ability to suppress neural signals to avoid "memory overflows" but without losing important inputs and to handle a mixture of neural spiking activity with and without information.

An interesting example is the visual system. When the lighting is gradually reduced the visual receptor cell input changes from cones to rods, demanding for a quite different signal management – but usually we are not aware that our visual information changes from colorful to black and white. Another important method is the handling of noise to detect weak signal, known as stochastic resonance [7]. Stochastic resonance is essentially a statistical phenomenon resulting from an effect of noise on information transfer and processing that is observed in both man-made and naturally occurring nonlinear systems [8].

12.4 Noise-Enhanced Auditory Information

At the threshold of hearing, the auditory system is able to detect vibrations of the basilar membrane with maximum amplitudes smaller than 10^{-10} m. Stochastic stereociliary movement caused by Brownian motions is with 3.5 nm [9] about 50

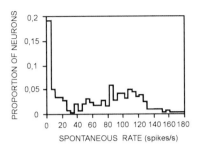

Fig. 12.4 Distribution of spontaneous firing rates in the auditory nerve of chinchillas. Three groups of fibers can be detected: fibers with (*1*) no or low spontaneous rate (19% of the fibers have less than 5 spikes/s), (*2*) medium, and (*3*) high (>40 spikes/s) spontaneous rate. The fibers of group 3 are able to detect low-level acoustic signals [11]

times stronger than the deterministic influence from the acoustic threshold signal. The Brownian motion is partly responsible for a phenomenon called "spontaneous activity": without any acoustic stimulus firing rates up to 160 spikes/s have been measured in auditory nerve fibers. The spontaneous firing rate of a nerve is strongly related to its sensitivity: the group of fibers without spontaneous activities are not stimulated by low-level signals at all [10]. However, fibers with high spontaneous activity transport the information of very weak stimuli by evolving regularities in their interspike times (time intervals between spikes). Nearly two-third of all auditory nerve fibers have high spontaneous activities with more than 40 spikes/s in silence [11] (Fig. 12.4). Therefore, in a healthy human ear, about 20,000 fibers are expected to react this way to low-level signals. However, due to the tonotopic principle, weak sinusoidal signals will cause only some hundreds of fibers to respond to the input signal and the number of neurons answering to an even weaker acoustic stimulus will further decrease. Our research group has simulated both the stereociliary movement as reaction to the Brownian motion of the inner ear fluids [12] and the IHC voltage changes caused by stereociliary motions [1]. It was shown that the contribution of Brownian motion can be approximated by Gaussian noise, low pass filtered mainly by the influences of the resistance and the capacity of the hair cell membrane.

In contrast to the hypothesis that noise limits perception [13], it is shown that for a signal which is too weak to reach threshold level, neural signaling is possible when noise is added. Such noisy signals result from thermal Brownian motion as well as by active forces from outer hair cells (Fig. 12.5). However, it is an additional task for the neural system to check whether an external signal is contained in the noisy neural input, and for positive cases to become aware of the signal features. The signal extraction method is based on the higher spiking probability during those periods where the positive part of the tone contributes to enlargements of the noise + tone signal.

The next examples show simplified computer simulations of the first 5 ms of basilar membrane motion when a 1000 Hz tone is applied. Based on finite element

12 Improving Hearing Performance Using Natural Auditory Coding Strategies

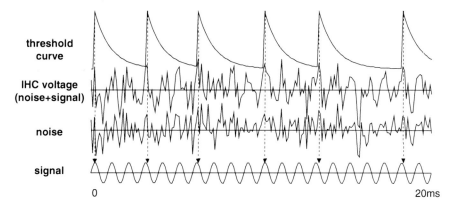

Fig. 12.5 Simulated voltage changes in an inner hair cell caused by a 500-Hz sinus tone and by the additional influence of noise (zigzag curve corresponds to the sinus + noise signal). An exponentially shaped "threshold curve" is shown which is used for calculating the firing times in a specific auditory nerve fiber: the threshold curve goes down to a minimum threshold value of about 0.1 mV. If IHC voltage change crosses this value, neurotransmitter release can occur, but afterward, for a special time (recovery period), spiking is rendered in a way described by the threshold curve. Note that the sinusoidal signal is too small to reach the threshold curve. Only with the help of noise, it can be detected in the neural pattern. Amplitude of the sinusoidal signal is 0.02 mV, rms (round mean square) amplitude of noise: 0.2 mV resulting in a signal-to-noise ratio of 0.1

simulations of inner ear forces, we assume that the tip of the longest inner hair cell stereocilia vibrate with an amplitude similar in size with that of the basilar membrane at the same position [14]. According to Fig. 12.6, spiking is therefore expected at places as marked by flash arrows in noiseless environment.

By adding noise as done in Fig. 12.5, the signal seems to disappear (Fig. 12.7). The situation becomes clearer as soon as more sensor elements are involved (Fig. 12.8). The source signal is now easy to recognize, from both the tonotopic and the temporal pattern: The frequency is recognized from the position with maximum regularity (place theory) as well as from the periodicity in spiking pattern (temporal theory) seen as dark stripes with 1 mm distances in Fig. 12.8. Under no-noise condition, a basilar membrane region with about 120 IHCs reaches the 2 nm threshold. This region is marked by two horizontal lines in Fig. 12.8. Remarkably the neural pattern becomes organized also outside of this supra-threshold band. In the next task, signal intensity is reduced by a factor 5 leading to a signal-to-noise ratio of 0.33 (Fig. 12.9). The 5 ms window again has now too few data for a definite decision. However, as most acoustical signals have a longer duration, we use now the next ten periods of 5 ms and we find in the superposition of this 50 ms data (Fig. 12.9, right) regularities even when covering about the upper half of the picture with a piece of paper. Shifting the paper in vertical direction gives an estimate for the detectable intensity corresponding to the threshold of hearing of a 50 ms, 1000 Hz tone.

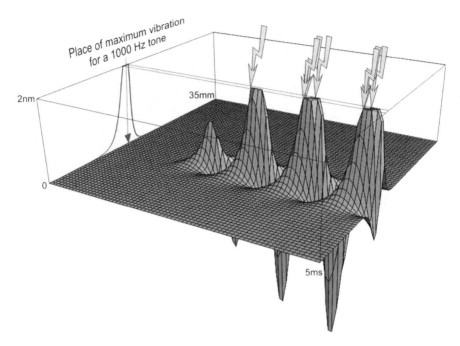

Fig. 12.6 Basilar membrane motion as function of time (0–5 ms) and space (0–35 mm) from a simple mechanical cochlear model. A weak pure 1,000 Hz tone causes vibrations in the 35-mm long centerline of the uncoiled basilar membrane, which finally has a region that reaches periodically a threshold limit of 2 nm. According to the tonotopic principle, the strongest vibration is at a specific position marked by an *arrow* on the left side. As indicated by the *curves* close to this *arrow*, amplitudes rise and fall exponentially as functions of basilar membrane's length coordinate. Decrease is quicker than increase, and this is the reason why low frequencies cannot be detected (quick signal lost) at positions a few millimeter after the "point of resonance" (right side upward from the *arrow* = apical direction)

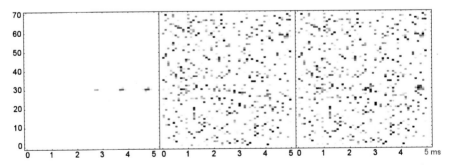

Fig. 12.7 Simulated neurotransmitter release from 70 inner hair cells that are equally distributed along the 35 mm length of basilar membrane. *Gray levels* indicate the basilar membrane maximum amplitudes for the 70 lines as functions of time, *darkness* corresponds with amplitude size – *white* means 2 nm are not reached. *Left*: signal without noise, *center*: noise without signal, *right*: signal + noise. All 70 hair cells cause spikes in the auditory nerve as marked by the *gray rectangles*, but the common input signal is lost in noise

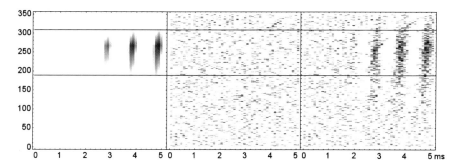

Fig. 12.8 Neurograms of the auditory nerve (spiking pattern) as pictures of neurotransmitter release from all the 350 inner hair cells that are within an "octave region" containing maximum vibration. Same method as in Fig. 12.7. *Left*: signal without noise, *center*: noise without signal, *right*: signal + noise

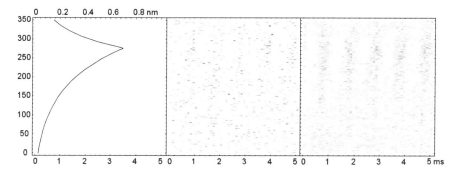

Fig. 12.9 Same situation as in Fig. 12.8 but for a five times smaller acoustical input signal. *Left*: signal without noise is empty, the place is used by a line showing the basilar membrane maximum vibration amplitude for 350 positions of IHCs with a highest value of 0.7 nm. *Center*: signal + noise; an order in the pattern is difficult to recognize. *Right*: superposition of ten 5 ms windows makes the order easy to see

12.5 Auditory Neural Network Sensitivity Can be Tested with Artificial Neural Networks

In a rather silent environment, we receive a lot of spontaneous activity and permanently the auditory part of the brain has to find out whether the auditory nerve input delivers just noise or noise with signal. Note that the signal can be a pure tone, a combination of tones, speech, or even noise. Just now, we have treated this task visually, that is by solving a related image recognition problem. We know that the visual system is rather good for such a job. Interestingly, artificial neural network techniques are also rather successful for image recognition tasks, and therefore we tested this tool to see what our brain can do in a similar situation using the biological neural network [15]. We found, e.g., that sinusoidal stimuli with a signal-to-noise ratio as low as 1/10 can be recognized from the simulated firing pattern of a single auditory nerve fiber. This seems to be rather a theoretical result as a 20 s data set is needed, and we cannot assume to store the single fiber data for such a long duration biologically. However, the same data will be selected by 100 fibers with high spontaneous activities from the same region in 200 ms which is more realistic. One should be aware about a quadratic relation for the sensitivity when all data are processed. This means we need to analyze a 200-ms interval from 400 fibers to reduce threshold of hearing by a factor of 2, resulting in a signal noise ratio of 1/20. Developing this technique, nature had to find a balance between number of fibers, their post-processing units, and the sensitivity of hearing. It seems to be a good choice that every inner hair cell has connections to several auditory nerve fibers, and most of them have spontaneous activities to support the "stochastic resonance phenomenon."

12.6 Cochlear Implants Versus Natural Hearing

After the pace maker for the heart, cochlear implants are the most successful devices for electrical nerve or muscle stimulation. It is surprising that auditory perceptions with these devices are rather acceptable in spite of the fact that they produce a quite unnatural spiking pattern in the auditory nerve fibers. The main difference to natural pattern is a strong synchronization of spiking times in large populations of fibers. With electrical stimulation, it is possible to obtain a bit higher spiking activities in the single nerve fibers, but the advantage of the volley principle (Fig. 12.2) is lost. The individuality of signaling using nearly 30,000 data lines in parallel is missed, and this is a pity as the richness of the neural auditory pattern is needed for a high fidelity quality in acoustic perception. Excellent development was done in miniaturization to obtain small devices which can be hidden behind or even in the ear. Many cochlear implant users have a bad speech understanding even under best acoustical conditions.

An essential restriction in obtaining more natural firing patterns results from the stimulation strategies generally used in human cochlear implants: all modern implants are purely based on the place principle. A common method is to activate in a cyclic manner up to 22 stimulating electrodes which are distributed along the cochlea: Their individual stimulus intensity locally activate the cochlear nerve according to the spectral characteristics of the auditory signal. But instead of additional support with temporal auditory information, the cycle period generates a virtual constant nonnatural temporal rhythm which is conducted by all stimulated fibers.

In relation with the third biological coding principle, several authors have investigated in adding noisy signals to the stimulus [16, 17]. A shortcoming of this approach is, however, that the added noise favors phase locking to the stronger signal parts of the noise and sustains artificially noise-related perceptions rather than really support the volley principle.

In general, electrical stimulation with cochlear implants obeys to the "all or nothing" low: either a spike is generated in an auditory nerve fiber or not. This principle causes phase locking between stimulus and response, and therefore much sharper poststimulus histograms for electrical than acoustical stimulation. The addition of noise is of some help but cannot achieve the natural distribution of fiber activation. However, a high-frequency background signal disturbs the synchronized firing within populations of neighbored fibers.

Therefore, a better method to enhance the temporal fine structure according to the volley principle seems the constant stimulation with a high-frequency signal with an intensity close to the threshold of fibers in the vicinity of the electrode. This signal alone will generate some stochastic individual firing that interrupts common phase-locked response rhythms as a consequence of refractory properties of the already pseudo-spontaneous spiking fibers. The phenomenon was discovered and analyzed to some extent by Rattay [18–20] and others [21–23].

12.7 Discussion

Our knowledge about the neural coding principles in mammalian auditory nerve fibers is primarily based on animal experiments. Single cell recordings enlightened our understanding how an acoustical signal is represented in the spiking pattern of the auditory nerve [6, 10, 24–27]. As the main elements of mammalian cochleae are quite similar and because of the restrictions for gathering human data, it is generally assumed that the same firing behavior can be expected in man. However, in the somatic region afferent human cochlear neurons are quite unique [28]. First, most neurons are not shielded by myelinated in this region, and second, many of them are gathered to clusters with two to four [29] neurons having a common insulation by myelin.

This morphological difference is of major relevance for the propagation of an action potential (AP) in the healthy ear and also responsible for quite different excitation patterns in case of cochlear implants. Both human particularities are expected to affect the neural pattern essentially, resulting in a specific human physiologic hearing performance. A first analysis of the electrical features of a non- or poorly myelinated somatic region demonstrated that the human afferent cochlear neuron is essentially lesser robust in spike conduction as that of cat and guinea pig, the preferred experimental animals [30]. Is loss of myelin in the somatic region a human imperfection caused perhaps by a genetic defect? A larger delay and a reduction in sensory information by loss of spikes seem to result in disadvantages only. Tylstedt and Rask-Andersen [31] speculate whether unique formations between human spiral ganglion cells may constitute interactive transmission pathways. These may be in the low-frequency region and may increase plasticity and signal acuity related to the coding of speech.

12.8 Conclusion

It is surprising that a combination of inner ear mechanics and a sensor cell type with a rather small operating range allows the high-quality detection of acoustic signals within a range of more than six orders. Impressing is furthermore how those noisy elements which cannot be eliminated, like Brownian motion of hair cell stereocilia, are used by nature for signal amplification in one of three quite different signaling strategies. When we recognize anatomical curiosities, e.g., the clustering of cell bodies in the auditory nerve which is unique in man, we should try to understand the neurophysiologic consequences, especially when we develop neuroprostheses that have to replace the sensory input. On one hand, it is unbelievable that artificially created neural pattern with a lot of nonnatural characteristics results at least in a low quality hearing in deaf people. But the challenge is to find solutions that add more of the natural features into the artificially evoked patterns.

References

1. F. Rattay, I.C. Gebeshuber, A.H. Gitter, The mammalian auditory hair cell: a simple electric circuit model. J. Acoust. Soc. Am. **103**(3), 1558–1565 (1998)
2. E.G. Wever, *Theory of Hearing* (Wiley, New York, 1949)
3. H. Helmholtz, *Die Lehre von den Tonempfindungen als Physiologische Grundlage für die Theorie der Musik* (Vieweg, Braunschweig, 1863) [English translation: A.J. Ellis, *On the Sensations of Tones as a Physiological Basis for the Theory of Music* (Longmans Green, London, 1875)]
4. W. Rutherford, A new theory of hearing. J. Anat. Physiol. **21**, 166–168 (1886)
5. E.S. Hochmair, I.J. Hochmair-Desoyer, Percepts elicited by different speech-coding strategies, in *Cochlear Prostheses*, vol 405, ed. by C.W. Parkins, S.W. Anderson, 268–279 (1983) (Ann. NY Acad. Sci.)
6. F. Rattay, P. Lutter, Speech sound representation in the auditory nerve: computer simulation studies on inner ear mechanisms. Z. Angew. Math. Mech. **77**(12), 935–943 (1997)
7. F. Moss, L.M. Ward, W.G. Sannita, Stochastic resonance and sensory information processing: a tutorial and review of application. Clin. Neurophysiol. **115**, 267–281 (2004)
8. K. Wiesenfeld, F. Moss, Stochastic resonance and the benefits of noise: from ice ages to crayfish and SQUIDs. Nature **373**, 33–36 (1995)
9. W. Denk, W.W. Webb, A.J. Hudspeth, Mechanical properties of sensory hair bundles are reflected in their Brownian motion measured with a laser interferometer. Proc. Natl. Acad. Sci. USA **86**, 5371–5375 (1989)
10. M.B. Sachs, P.J. Abbas, Rate versus level functions of auditory nerve fibers in cats: Tone burst stimuli. J. Acoust. Soc. Am. **56**, 1835–1847 (1974)
11. E.M. Relkin, J.R. Doucet, Recovery from prior stimulation. I: relationship to spontaneous firing rates of primary auditory neurons. Hear. Res. **55**, 215–222 (1991)
12. W.A. Svrcek-Seiler, I.C. Gebeshuber, F. Rattay, T.S. Biro, H. Markum, Micromechanical models for the Brownian motion of hair cell stereocilia. J. Theor. Biol. **193**(4), 623–630 (1998)
13. W. Bialek, Quantum noise and the threshold of hearing. Phys. Rev. Lett. **54**, 725–728 (1985)
14. W. Müller, *Untersuchung der Nachschwingzeit in der Cochlea unter Berücksichtigung der Reissnerschen Membran (in German)*, Master Thesis, TU Vienna, 1996

15. F. Rattay, A. Mladenka, J. Pontes Pinto, Classifying auditory nerve patterns with neural nets: a modeling study with low level signals. Simul. Pract Theory **6**, 493–503 (1998)
16. R.P. Morse, E.F. Evans, Enhancement of vowel encoding for cochlear implants by addition of noise. Nat. Med. **2**, 928–932 (1996)
17. M. Chatterjee, M.E. Robert, Noise enhances modulation sensitivity in cochlear implant listeners: stochastic resonance in a prosthetic sensory system? J. Assoc. Res. Otolaryngol. **2**(2), 159–171 (2001)
18. F. Rattay, High frequency electrostimulation of excitable cells. J. Theor. Biol. **123**, 45–54 (1986)
19. F. Rattay, *Electrical Nerve Stimulation: Theory, Experiments and Applications* (Springer Wien, New York, 1990)
20. F. Rattay, Basics of hearing theory and noise in cochlear implants. Chaos Solitons Fractals **11**, 1875–1884 (2000)
21. L. Litvak, B. Delgutte, D. Eddington, Auditory nerve fiber responses to electric stimulation: modulated and unmodulated pulse trains. J. Acoustic. Soc. Am. **110**(1), 368–379 (2001)
22. R.S. Hong, J.T. Rubinstein, Conditioning pulse trains in cochlear implants: effects on loudness growth. Otol. Neurotol. **27**(1), 50–56 (2006)
23. B.S. Wilson, R. Schatzer, E.A. Lopez-Poveda, X. Sun, D.T. Lawson, R.D. Wolford, Two new directions in speech processor design for cochlear implants. Ear. Hear. **26**(4), 73S–81S Suppl. S (2005)
24. N.Y.S. Kiang, *Discharge Pattern of Single Fibres in the Cat's Auditory Nerve* (MIT Press, Cambridge, 1965)
25. J.F. Brugge, D.J. Anderson, J.E. Hind, J.E. Rose, Time structure of discharges in single auditory nerve fibers of the squirrel monkey in response to complex periodic sounds. J. Neurophysiol. **32**(3), 386–401 (1969)
26. S.A. Shamma, Speech processing in the auditory system. I: the representation of speech sounds in the responses of the auditory nerve. J. Acoust. Soc. Am. **78**, 1612–1621 (1985)
27. E. Javel, Shapes of cat auditory nerve fiber tuning curves. Hear. Res. **81**(1–2), 167–188 (1994)
28. J.B. Nadol, Jr, Comparative anatomy of the cochlea and auditory nerve in mammals. Hear. Res. **34**, 253–266 (1988)
29. S. Tylstedt, A. Kinnefors, H. Rask-Andersen, Neural interaction in the human spiral ganglion: a TEM study. Acta Otolaryngol. **117**(4), 505–512 (1997)
30. F. Rattay, P. Lutter, H. Felix, A model of the electrically excited human cochlear neuron. I. Contribution of neural substructures to the generation and propagation of spikes. Hear. Res. **153**(1–2), 43–63 (2001)
31. S. Tylstedt, H. Rask-Andersen, A 3-D model of membrane specializations between human auditory spiral ganglion cells. J. Neurocytol. **30**(6), 465–473 (2001)

Index

Acrylates, 112, 113, 116
Activating function, 220
Adaptive systems, 136, 139
Adhesion, 25, 32, 36, 38, 40, 41, 43
Algorithmic design, 166
Analogy research, 129
Anisotropy, 85, 96
Architecture, 149–175
Architecture theory, 149
Architekturbionik, 127–146
Artificial neural networks, 229, 257–258
Attachment, 25, 32, 37–41
Auditory perception, 251, 258
Auditory system, 249, 253
Automation, 203–217
Avantgarde, 150, 152, 165–167, 174
Average condition, 86

Bearing curve, 70–75, 77
Bearing load, 64, 68, 70–73
Best practices, 25, 31–40
Biodegradable materials, 115, 116, 121
Bioinspired, 32, 34, 38–41
Bioinspired facade design, 141, 143
Biologically Inspired Design, 132, 143
Biology, 25–29, 39, 43
Biology for engineers, 28–30
Biomaterial, 85, 90, 94–97, 106
Biomimetic architecture, 134
Biomimetics, 25–43, 84
Biomimetics in architecture, 127–146
Biomimicry, 133, 134, 136, 145
Biomimicry Guild, 31
Biomimicry Innovation Method, 25, 31–42
Biomorphism, 145, 149–175
Bionic structure, 11, 13–22

BioScreen, 43
BioTriz, 133, 134, 136
Blending of disciplines, 25
Bone, 81–98
Box-counting dimension, 197, 198
Box-counting method, 198, 199
Brain computer interfaces (BCI), 230–231
Buckling, 32, 35–38
Building block, 84
Built environment, 145, 146

Castel del Monte, 189
Cationic polymerization, 113, 114
Cell membrane, 10, 12, 13, 16
Cell size, 120
Cellular material, 118–120
Cellular structures, 105–121
10-channel stimulator, 236, 237, 241
Chaos theory, 167
Closed loop, 227–230, 243
Coastline, 181–185, 190–192
Cochlea, 250–253, 258, 259
Cochlear implants, 251, 253, 258–259
Coherence, 182, 184, 186, 189, 195–197, 199
Collagen, 84, 88–94, 96, 97
Concentration tensor, 87
Continuum micromechanics, 84–87
Contraction dynamics, 225, 226
Control, 219, 222, 224–232, 235, 237, 238, 241–243
Corral reefs, 31
Correlation of form and function, 29
Cortical bone, 89, 91, 94–96
Crystal, 88, 90, 93, 96–98
Curdling, 192–195
Cyborg, 159–162

Data transmission, 216
Deconstruction, 165, 166
Descriptive, 26
Design (top-down), 212–213
Design characteristics, 96–98
Design methods, 150, 165
Design program, 132, 138, 140–144
Deterministic surface, 52, 56
Diatoms, 32, 34, 37, 38, 40, 41
Distance, 183, 184, 196–199
Dynamic mask based stereolithography, 108–110

Ecology, 162–165, 174
Ecosystems, 31, 33
Elasticity, 86, 90, 93, 95, 96
Elastoplasticity, 86
Electrode, 220, 222–224, 230, 242
Elementary component, 88–91
Emergence, 130, 134–135, 146
EMG, 229, 230
Energy efficiency, 136, 137
ENG, 229, 230
Ergometer rowing, 239–241
Euclidean modern architecture, 181, 184
Euplectella aspergillium, 107
Extracellular bone matrix, 89, 92, 93, 96
Extrafibrillar space, 88, 92, 93, 96
Extravascular bone matrix, 88, 89

Factor chance, 191–193
Fatigue, 32, 35–38
FES control, 230
FES cycling, 231–238, 241, 243
Fibril, 84, 88, 89, 92–94, 96
Field bus, 205–209
Finite element method, 118
Finite element modelling, 121
Flora, 33
Foot drop, 219, 241, 242
Force measurement cranks, 235–236
Forward dynamics, 227
Forward dynamic simulation, 232, 233
Fractal architecture, 192
Fractal geometry, 159, 165, 166, 179–199
Fractal-like, 182, 185, 186, 189, 192, 197, 199
Fractal properties, 182, 185, 192, 199
Friction, 25, 34, 43, 52–57, 62, 63, 72, 77
Fuel cell (FC), 14, 15, 18, 22, 23
Function, 27, 31, 33, 37, 38, 40, 152, 157, 162, 167–173

Functional electrical stimulation (FES), 219–244
Fused deposition modelling (FDM), 107

Gaudí, Antoni, 187
General principles, 29
Genetic algorithm, 227, 233
Goff, Bruce, 186
Gothic window, 187, 188, 192
Gramicidin-A (gA), 17

Hair cell, 249–256, 260
Hard sciences, 26
Hierarchy, 29
Hill tensor, 87
Homogenization scheme, 86, 91
Honing, 51, 53, 73, 74, 77
Human archetyp, 204
Hybrid control, 228
Hydroxyapatite, 84, 88–98

Influence tensor, 87
Information theory, 204–206
Infrared vision, 22
Initiation, 112
Inkjet based systems, 110–111
Innovation potential, 31, 33
Instrumented ergometer, 240–241
Integration, 136–138
Intelligence (artificial), 209–210
Intelligence (definition), 210–212
Intelligence (systems), 209–214
Intelligent building, 136
Interactivity, 135
Interdisciplinarity, 214
Interdisciplinary, 28
Interoperability, 205, 213
Inverse dynamics, 226, 228
Isokinetic, 232, 233, 238
Isometric recruitment curve (IRC), 222
Iterated function system (IFS), 180
Iteration, 182–191, 193–197

Kinematic chain, 239
Kinetic chain length, 116
Knowledge transfer, 25
Koch curve, 183–185, 191, 194, 197, 198
Koch island, 191

Lacunae, 88, 93, 94
Lateral-line organ, 249, 250

Le Corbusier, 181
Length measurement, 182–184
Light construction, 129
Lightweight design principle, 105
Localization, 119, 120
Locomotion, 54, 55, 57, 72, 73
Lubrication, 25, 28, 31, 33–35, 43, 52, 54, 55, 64, 73, 74, 77
Lynn, Greg, 189, 192

Malaysia, 33, 42
Material phase, 85, 86, 90
Matrix-inclusion problem, 87
Mechanical test, 90, 94
Mechanics, 84
Micro-architecture, 106
Microelectromechanical systems (MEMS), 34, 37
Micromachinery, 40
Microstructure, 85, 86, 88, 89, 92, 93
Mimetics, 152
Mobile cycling, 237–238
Model (behavioristic), 213
Model (choosing), 212
Model (hierarchical), 209–210
Mori–Tanaka scheme, 87, 91, 93
Morphogenesis, 134–135, 146
Mountain algorithm, 190
Multifunctionality, 33
Multiscale, 81–98
Muscle activation dynamics, 227

Nanomembrane, 9–23
Nanotechnology, 27, 28
Natural construction, 129–131
Natural laws, 26
Nature, 25, 27, 29, 31, 32, 34, 36–41, 149–153, 155–157, 159, 162, 165, 167, 168, 170, 172–175
Nature and fractals, 190
Neural code, 259
Neuroprosthesis, 220, 221, 223, 231, 243, 244, 260
Newton–Euler equations, 226, 232, 239

Observation scale, 88, 91, 93, 94
Open loop, 227, 228
Optimal stimulation pattern, 227, 228, 233, 237, 239
Optimization, 29, 30, 227, 233, 234, 239, 241
Organicism, 149, 150, 152, 173, 174

Organic unity, 152, 167, 171–174
Organisms, 25, 28, 30–32, 36–38
Organ of Corti, 252
Orthosis, 232, 236

Paradigm shift, 33
Perception, 184, 196
Photoinitiator, 112–115
Photopolymer, 108, 110–117, 121
Place theory, 253, 255
Polylactic acid, 115, 116
Polyvinyl alcohol, 117
Porosity, 94, 96, 97
Principles, 26, 27, 29, 42
3D-printing, 108
Processes, 25, 27, 29, 31–33, 42
Profiles, 205, 207
Propagation, 112
Proportion, 152, 167
Proton transport, 15–17
Psychoanalysis, 204, 214–217
Python, 55–58, 62, 63, 73–77

Radical polymerization, 112–114
Rainforests, 31, 33, 39
Representative volume element (RVE), 85–87, 93, 94, 96
Resources, 32
Response, 52, 53, 56
Rietveld, Gerrit, 186
Robie House-Frank Lloyd Wright, 181, 185, 198
Roughness, 182–184, 197, 198

Safety, 205, 208, 209
Scaffold, 106, 115, 116
Scaling, 165, 166
Scenario recognition, 216
Scientific publishing, 43
Security, 205, 208, 209
Selective laser sintering, 106
Self-consistent scheme, 87, 93
Self-similar dimension, 194, 195, 198
Self-similarity, 181, 182, 184–187, 189, 190, 192
Sensitivity analysis, 234
Shape change, 135
Shear, 32, 35, 36, 38
Simulation, 146
Simulations, 191–196
Sliding, 52, 54–56, 60–64, 71–74, 76, 77

Smart materials, 133, 137
Southeast Asia, 33, 39
Species variety, 31
Spontaneous activity, 254, 257
Squamata, 54
Squamate, 54, 58
Stationary cycling, 238
Stereocilia, 250, 255, 260
Stereolithography, 108–110, 112, 114, 121
Stiffness tensor, 87
Stimulation pattern, 225, 227, 228, 232–234, 236, 237, 239, 241
Stimulation signal, 220, 222–223
Stimulator, 219, 220, 222, 224–225, 236–238, 241, 243
Stochastic resonance, 253, 258
Strain, 86, 87, 93, 96
Strength, 84, 85, 88–91, 93–97
Stress, 85, 86, 90, 96
Structures, 32–36, 41
Superposition, 86
Surface roughness, 71, 77
Sustainability, 29
Systems, 25, 27, 29–31, 34, 42

Technology transfer, 29, 39
Technoscience, 27
Temporal theory, 253, 255
Termination, 112
Tetraplegics, 242
Thermal infrared detector, 20
Three gaps theory, 42, 43

Tourism, 33
Trabecular bone, 94, 95
Tractions, 55, 56, 62, 63, 72
Traditional building, 129
Tree algorithm, 191
Tribology, 25–43, 53–55, 73, 74
Tricycle, 231, 232, 235, 238, 239
Tropical rainforests, 31
TU BIONIK Center of Excellence for Biomimetics, 1, 25
Two photon polymerization (2PP), 111, 115, 121

Ultrasound, 100, 101, 103
Ultrastructure, 88, 89, 92, 93
Ultrathin films, 19
Uniaxial stress, 118
Upper limb, 242–243

Vernacular architecture, 130
Volley principle, 251, 258, 259
Volume fraction, 85, 86, 91, 93, 94

Walking, 229, 231, 242
Water, 84, 88–91, 94, 97
Wear, 25, 28, 29, 32–35, 38, 43, 54, 71, 73, 77
Wright, Frank Lloyd, 181, 185, 198, 199

Young's modulus, 118